TO MY SON HARRY, REMEMBERING THE

CHRISTMASES WE SPENT TOGETHER

A Treasury of Christmas Decorations

ZELDA WYATT SCHULKE

HEARTHSIDE PRESS, INC., *Publishers*

NEW YORK

Contents

Introduction to Christmas Decorations

Decoration is one of the principal activities of Christmas. We wrap gifts, trim trees, and bedeck our homes from mailbox to chimney-top. Stores, towns and cities erect fantastic displays to make us stop and stare. Everywhere is glitter and paint, tinsel, gold paper and gaily colored baubles that appear only at Christmas. All these things and many more make decorating for the holiday season an exuberant chore that fascinates designers in every field from flower arrangement to architecture. This is art, not for art's sake, but for fun and gaiety!

Such exuberant designing is a welcome relief to artists who turn to it from more sophisticated projects. These creative individuals develop decorations of real beauty that serve as inspiration for many to copy. Each year there is a demand for new ideas in Christmas decorations and information on how to make them. This book is dedicated to that demand. It is for families who like to make their own Christmas decorations and who are discriminating enough to want them to be beautiful. Here are decorations to fit in various types of homes in many regions of our country. Detailed information and easy *how-to* pictures will help you set the stage for a merry Christmas.

Christmas is a time when all the glitter and sparkle of the world, the bits of color and forgotten treasures, become a part of the beautiful picture of the season. So let us gather tinsel, buttons, sequins, ribbons, paper, plant materials, in a multitude of colors, and create out of them the spirit of Christmas. Be aware of ideas and materials all around you. The beauty and magic of the season never just happen—they are planned and worked for. You are your own interior decorator at this, the best time of year to practice the principles of design. Learn how effective the results can be and experience the joy and satisfaction of creating real beauty.

9

A Treasury of Christmas Decorations

ACKNOWLEDGMENTS

I wish to express my grateful appreciation and thanks to my family and my husband in particular, also to Kathryn Seibel, Mike and Ruth Bryan for their help and encouragement, and to all the flower arrangers who contributed pictures and knowledge to enrich this book.

Design in Christmas Decorations

You may wonder how an artist goes about materializing his ideas and developing beautiful decorations. He does not necessarily spend a great deal of money to create his effects. Good taste, imaginative use of materials, and respect for design principles are his forte. Be aware of materials all through the year, wherever you are, and collect them to use at Christmas. Collect dried and treated foliage, seed pods and cones from the garden, the wayside and the woods, and artificial material found in gift shops, variety stores and florists. You may find many treasures at white elephant sales and in secondhand stores.

Any well designed decoration, regardless of how complicated or spectacular, is the result of blending the basic elements of design (line, form, space, light, color and texture) into a unified whole. Each design unit must declare a clear, direct and strong visual statement, complete as such.

The principles of design (proportion, scale, balance, unity, contrast, dominance and rhythm) are the guides that help the designer develop decorations of beauty. As you make your Christmas decorations, your job is to develop order, interest and beauty. Order will be achieved by understanding and applying the principles of design to your work. Understanding these principles is a lifetime study, but these simple definitions may help to clarify their meaning.

PROPORTION is the size relationship of one element of a design to another and to the space it occupies. The use of proportioned lines, forms or colors, helps to establish a feeling of rhythm and unity, binding all the elements together into a whole so tightly, that removing any single part would disturb the whole design.

SCALE is the size relationship of materials. Use a sequence of sizes and shapes in your design.

BALANCE is an inherent need in our make-up. We look for it in all visual objects. In a design it is a matter of weight, visual and actual. The right space between elements of a design establishes balance. By changing the fulcrum a more interesting balance is

often achieved. For unity and harmony, the lateral, horizontal and depth balance must be correct. Let your inborn feeling for balance guide you as you design.

UNITY requires that all the elements of a design be woven together according to a well laid plan. Draw a plan on paper, or keep it in your mind's eye, for you must have one to bring completeness and harmony to a composition.

CONTRAST, sometimes called variety, is just as important as harmony, for it brings interest to a design. By contrasting colors, lines, shapes and textures, stimulus and interest are produced and monotony is avoided. Round acorns and straight pine needles are contrasts in shape. Red and green are contrasts in color. A smooth satin bow is a contrast for a straw basket.

DOMINANCE requires that one line, one form, one color, one texture dominate the design, the other elements must be subordinated. As you make a decoration, conflicts occur in line, form and color. Dominance resolves these conflicts and restores unity. Without it a design tends to disintegrate and rhythm may be destroyed.

RHYTHM is progressive measured action. It is a graceful movement of lines, forms, colors and textures that carries the eye easily through the design. It is attained by the use of repetition, transition, gradation, spacing and color blending of the elements. Rhythm is a basic aesthetic need. It is something we feel and sense in all of life and is an important art principle and the spectator immediately recognizes and responds to it.

By applying these principles of design, the elements that we work with (line, form, space, light, color and texture) are bound together to produce designs of beauty.

All aspects of design play an important part in making each decoration; but color is the element that enjoys the spotlight at Christmas. Over the years, almost every family collects treasures. Madonnas, angels, crèches, stars, Santas, reindeer and bells become a part of the Christmas tradition. Perhaps each member of the family has a different favorite, and we can use them all if we unify the entire decorating plan with a dominant color harmony. Choose a harmony that is right for the established colors in your home and use it on the door, the mantel, the tree, the dining table and throughout the house.

For many years the traditional Christmas colors were red and green, and today red and green still spell Christmas for many people. If this is true with you and if the established colors in your rooms can take it, by all means use red and green, for being true complements, they are visually satisfying.

With the vogue for pastel colors in home décor, the practice of choosing a gay new color harmony for Christmas has developed. Well chosen colors that harmonize or contrast with the room, can say Christmas just as surely as the traditional red and green. Choose a color scheme that suits your home and your personality, but vary it from year to year to give a new look to your Christmas treasures. The sources of color inspiration are many and varied, but fashion is the dictator. What color is popular this year? Can you use it in your home? If so, do, for it will add a note of up-to-dateness to your decorations. A contemporary trend is to move in the same hue from low to high intensity. Dark green evergreens used with brilliant yellow-green Christmas balls, for example. Another is to keep in one color range and then switch to a small amount of a contrasting color for punch. Try a della Robbia wreath done in pinks and violets with a spot of bright copper-orange for accent. One color area can make another advance or recede. Intense color can be held back by using its grayed complement with it. Or color can be held back by carrying these intense colors down to their own darkest values or lowest intensities.

Colors affect emotion. Warm colors of low intensity—maroons and dark reds, for example—make you feel good but do not excite you. If you want excitement, use analogous colors, going over to the second primary color or farther. (Red, red orange, yellow orange and yellow, for example.) If you go far enough you can create a feeling of carnival.

Color is endowed with a friendly persuasion. Hues, values, chromas can set the mood of a decoration. By combining colors, values and intensities you can give all grades of feeling to your work, for color makes the most emphatic impact on the spectator. Generally speaking we may say that yellow, orange and red suggest warmth, friendliness and gaiety. Blue, violet and green suggest coolness, reserve, dignity and, in the case of purple, royalty perhaps. Colors are pleasing because they are similar (related) or because they are unlike (contrasted). Related colors have no dead ends or stops, they merge easily one into the other.

All the design principles apply to the handling of color just as surely as they do to the handling of line and form. A vivid color at the focal area attracts attention as forcibly as a dramatic line or form. Develop an awareness of colors and forms around you, decide which belong together and which are best related to the space where you will use them. As you work with color, remember that *the more varied the colors used, the more difficult it is to control the design.* Color rhythm brings unity and interest to a design through planned relationships, through the orderly transition and gradation of hues, values and chromas.

Dominance is perhaps the most valuable principle to aid in the handling of color. Dominance suggests that one hue, value or intensity should rule in a composition. The hue of largest area or dominance sets the color key. Dominance can restore unity to a composition when too great conflict in color occurs.

After all is said that can be said about color, it is the individual handling of color that makes a decoration truly personal and your ultimate aim should be to express your own taste.

A Warm Welcome at the Door

Do you share Christmas with your neighbors and all who pass your way with gaily decorated doorways, mailboxes, fences, bird feeders and lamp posts? Many do. Each year an increasing number of well designed outdoor decorations appear and are gradually supplanting the overdone, multicolored lights that for years have been draped helter-skelter over doors, windows, roof-tops and foundation plantings in a show of poor taste.

Outdoor Christmas lighting can be beautiful if it is planned by someone who knows how to use light and color. In a small village in Ohio garden club members were asked to submit plans for town decorations to the local businessmen's group who had always handled it. They submitted a plan for an *all white Christmas,* and to their surprise it was accepted. Thousands of white lights, and white lights only, decorated the Christmas trees that lined the

1. *Designed by:* Kathryn Holley Seibel
Photographer: Bryan Studio

main highway in front of every store, church and civic building and in the village square. White lights shone from the belfries of the churches and the Town Hall that faces the square. Christmas Eve a blanket of snow came down over the village and it was like a glimpse into fairyland. Travelers from many states stopped to comment on its beauty.

Most of us have seen in pictures at least the beautiful Christmas lighting of Rockefeller Plaza in New York, and every traveler who crosses Pennsylvania during the Christmas season looks for the star over the city of Bethlehem.

Don't Look Now, But Your Mailbox is Showing, was the slogan for a Christmas decorating contest for mailboxes in a rural community. Many beautiful and delightfully imaginative decorations appeared, adding gaiety and interest to the village. Plate 1 was one of the prize winning boxes. The name and the musical notes were painted bright red. The rooster was made from wheat straw and its red plastic comb repeated the red lettering. Not everyone could make such a rooster, but the possibilities of appropriate and exciting decorations are unlimited. An evergreen and cone swag tied with a bright red weatherproof bow, Santa and his reindeer riding over the roof of the box, are just two possibilities. Rural mailboxes are often eyesores on the horizon, but with a bit of Christmas whimsy they can add charm to the holiday picture. ·

In the southern part of the United States and on the west coast, tropical and semitropical plant materials are used for Christmas decorations with outstanding success. We get into the habit of looking at the plant forms we see every day without really seeing them or realizing their potential. Appreciate the plant life of your region and rediscover its beauty and design.

Miss Reba Harris of Eustis, Florida, has done much to make northern Floridians aware of the great possibilities in native plant material. Following is a quotation from a pamphlet that she wrote, *Let's Have A Real Florida Christmas:*

> "Palm trees graced the manger scene
> On that day long gone by
> When Wise Men came to Bethlehem
> And angels sang on high.

> This verse discovered on a Christmas card, gave us an idea that opened up a joyous approach to Christmas decorations in Florida. Why of course, the first Christmas tree was a palm! The Holy One

2. *Designed by:* Reba Harris   *Photographer:* John R. Walter

was born where the palm trees grow—and Florida the land of the friendly nodding palms, with a climate like that of the Holy Land, has the real Christmas tree. Why not develop in Florida a Christmas more in harmony with our native plants and trees, our warm sunny days, with ripening fruit and flowers everywhere? Why try to imitate a northern Christmas? Let's deck the small palms with gay ornaments for tables and windows indoors; large ones in yards and gardens and public parks with floodlights or the traditional Christmas tree lights. Let's use native plants, more flowers and parts of palm, sprayed silver or gold for house decorations. Let's feature the palm in our outdoor Nativity scenes."

This idea met with immediate response. Children and adults alike adopted the palm as Florida's Christmas tree and use it in many of their decorations, together with other native plant material. (See Plates 2, 12 and 13.)

Many more examples of fine civic holiday lighting and decorating could be cited, but these are the exceptions, and a great deal of constructive work remains to be done. As in every good endeavor, the place to start is at home. Each one of us can see to it that our doorway radiates a well designed, cheery holiday greeting for those who pass our way.

3. *Designed by:* R. G. and H. W. Schulke
Photographer: Bryan Studio

In Plate 3, white pine roping and a large cone and evergreen wreath decorate a high sapling fence. These fences are popular today and provide a desired privacy but may seem a bit austere and forbidding. The festive decorations soften them and seem to radiate a feeling of good will and hospitality. The pine roping may be purchased from your florist. The cone wreath was made on a 28-inch 4-wire frame, using the same method as shown in Plate 17.

4. *Designed by:* Barbara S. Meisse and Carlton B. Lees
Photographer: Earl E. Roberts

Plate 4 is a close-up view of the beautifully executed swags that decorated the entrance pillars at Kingwood, the new horticulture and civic center in Mansfield, Ohio. The panel, 6 feet long, is made on a wooden, rectangular frame. Gilded laurel foliage was wired to the frame for the foundation and white pine for the center of the panel. Large California sugar pine cones were sliced lengthwise, superimposed on the pine and wired to the wooden frame. The cone roses are made by slicing these same cones crosswise. Three of these cone roses, together with green Christmas balls complete the focal area of these handsome gate panels.

The outside door is the place to establish the color harmony and the type of materials that will be used in all the decorations throughout the house. In a house where there are small children, Santa Claus, his helpers, reindeer, snowmen, toys and candy might well be chosen for the dominant theme.

The clown with toy horn and balloons, on the front door in Plate 14, the candy tree in Plate 83, candy canes, toys and Santa on the big Christmas tree, snow people and snowballs on the mantel as in Plate 44, are good examples of recurring themes that bring unity and charm to a decorating plan.

The della Robbia theme in Christmas decorations has grown in popularity in recent years. Artificial fruit in muted colors is now imported from Europe and is available in department stores, variety stores and at the florists. This fruit lasts for years, making it a wise investment. Used alone or combined with natural material, an almost endless variety of elegant and dignified decorations can be made. See Plates 18, 21, 28, 35, 71, and 78 for the recurring della Robbia theme.

Natural plant material remains the all-time favorite for Christmas decorations. One tradition says that greens in the house provide branches for the good spirits of Christmas to dance on. In the northern part of the United States conifer and broadleaf evergreens with cones and seed pods are combined to make decorations of real beauty, unified through the use of a dominant color scheme. See Plates 3, 5, 10, 11, 29, 44 and 75.

In the southern part of our country and on the west coast, tropical and semitropical plant materials are used with outstanding success to fit the homes of those regions. See Plates 12, 13, 22, 60, and 63.

In Plate 5, a heart-shaped boxwood wreath and topiary trees at the door of this white frame house set the theme and the color harmony for all the inside Christmas decorations. The mantel of this house may be seen in Plate 47.

5. *Designed by:* Mrs. William Siemon
Photographer: William E. Buvinger

To make this door decoration, spray a heart-shaped shallow basket with gold paint. Wire a gold cherub to the basket. Fill a 12-inch 4-wire frame with wet sphagnum moss, covered with aluminum foil as in Plate 16. Fill with boxwood to conform to the heart shape of the basket. Wire the basket to the frame. Attach a bow with streamers of ombre satin, shading from pink to deep scarlet to the bottom of the door.

To make the topiary boxwood trees shown at each side of the door, fill a 9-inch chicken wire ball with wet sphagnum moss. Attach balls to two 1-inch oak dowels 3 feet long. Wind the dowels with brown floral tape to make the trunks of the trees. Cut boxwood twigs in 9- to 12-inch lengths and thrust into the wet moss until the ball is completely covered. Decorate the trees with pink Christmas balls and crab apples painted gold and sprinkled with glitter. Put the trees in 12-inch clay flower pots painted dark green. Pour plaster of Paris into the pots until they are three-fourths full. For a finished effect, fill the rest of the pot with peat moss. The wreath and the two trees require 7 pounds of boxwood. The roping around the door is made by cutting Scotch pine in 6-inch lengths and wiring them to a clothesline of the required length to fit the door frame.

Storm doors present a design problem in the northern United States. In the engaging Christmas door in Plate 6 the cross lines of the windowpane have been used to suggest the fence over which Santa's reindeer jumps—a bit of wise, imaginative planning. This is an aqua door on a pink house. The reindeer is made of pink cardboard with glitter and pink balls on the horns, and glitter outlining the body. To make the swags on the corners of the door, wire two pieces of juniper together in crescent shape. Spray with clear lacquer and sprinkle with silver glitter while the lacquer is wet. Paint the pine cones pink and sprinkle with silver glitter. Wire the cones to the juniper. A floodlight concealed in the foundation planting illuminates this colorful doorway.

The door swag in Plate 7 is made on a coat hanger frame. Pull a coat hanger into diamond shape, with the hook at the top. Cover frame with chicken wire. As you make the swag, put the end of each evergreen branch into one hole of the chicken wire, bring the stem of the branch under the chicken wire and up into another

6. *Designed by:* Mrs. Carl Schmalstig
Photographer: Commercial Photos, Inc.

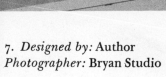

7. *Designed by:* Author
Photographer: Bryan Studio

hole, as in Plate 8. This will hold the branch secure and makes for speed and ease in designing. To duplicate this swag use flocked balsam branches, put them in the chicken wire in a good line pattern of radiating design. At a variety store purchase a white, glittered half bell of papier mâché for the center of interest. Decorate the bell with a group of pink Christmas balls, pink glitter daisies, and a pink metallic bow with streamers.

Many similar swags can be made to fit your particular need. Let the center of interest carry out your theme and color harmony. If you have small children in your family, try using some of their toys for the center of interest and watch their eyes shine with delight.

8. *Photographer:* Bryan Studio

9. *Designed by:* Mrs. Edward Johnson   *Photographer:* Bryan Studio

In Plate 9, a handsome festoon of cones, seed pods and taxus makes an impressive door decoration. The swag on the iron railing below carries the eye down to the foundation planting, unifying the composition. To make a cone festoon, see Plate 42.

Cut a piece of chicken wire the length of the entire festoon. The center section of this one is 4 feet long, the side pieces 3 feet long, requiring a piece of chicken wire 10 feet long. Roll the wire into a cylinder, flatten it and shape into festoon form, pinching the wire together at the two upper corners. Wire the cones, seed pods and small bunches of taxus separately and attach these to the chicken wire frame. Use slender seed pods at the ends of the festoon to give a diminishing line to the design. Bows to harmonize with your color scheme may be made and wired to the corners of the festoon.

10. *Designed by:* Mrs. Richard Barnes   *Photographer:* Bryan Studio

Cone festoons make beautiful indoor as well as outdoor decorations. This elegant cone festoon, Plate 10, was designed for the stairway of a Western Reserve farmhouse. To duplicate it, cut a piece of heavy hemp rope the length required for your staircase. Wire the cones and seed pods separately and wire these to the rope in continuous design. Long lasting evergreens, such as taxus, white pine or cedar may be added and they will keep through the holidays. The cones, of course, will last for years and it is a simple matter to add fresh evergreens to the festoon each year.

11. *Designed by:* Mrs. Aubrey Holladay
Photographer: Bryan Studio

Plate 11 is another variation of a cone festoon that is very easy to make. Buy white pine roping from your florist the length required for your staircase. Wire cones and seed pods separately and then wire these to the roping. Or wire them to hemp roping, according to the directions given under Plate 9 and then back it with the pine roping. By using the latter method, the cone festoon is ready to use for many years. Pine roping was also used on the mantel of this Georgian house, Plate 37, bringing unity to the decorating plan.

12. *Designed by:* Mrs. W. G. Wells
Photographer: Glenn H. Bolles

The smartly-styled swag, in Plate 12 was designed for a Florida
door. It is made of cut Sabal palm leaves, dry teasels, gold and red
Christmas balls and a red satin bow. To duplicate this, trim two
Sabal palm leaves, pointed at the top and rounded at the stem.
Wire the two together, one up and one pointing down. Spray with
gold paint. Paint teasels gold and wire to the swag. Wire two
bunches of red and gold Christmas balls, that may be purchased
in grapelike clusters, to the palm, and also the bright red satin
bow. The focal area and the colors may be varied to suit your particular theme and color harmony.

Semitropical plant material is used in the unusual door decoration, Plate 13. Split the edge of a very large, fresh palmetto palm leaf, at every fold. Roll the cut pieces on a pencil and secure each roll with a bobby pin. In a few days, when the palm leaf has dried, remove the bobby pins and pull the rolls out into curls. Spray the palm with gold paint. The bell-like decoration is made by inverting the palm epergne shown in Plate 63. Wire a red Christmas ball to each bell for a clapper. To make the bow use fresh honey locust seed pods, bend into loops and wire together. These pods are a beautiful chartreuse color when fresh, completing a pleasing color scheme of gold, red and chartreuse. Or spray the honey locust bow with bright red lacquer for a gayer effect.

13. *Designed by:* Author
Photographer: Bryan Studio

14. *Designed by:* Author   *Photographer:* Bryan Studio

In Plate 14, a gaily painted papier mâché clown, one of Santa's helpers, with horn and balloons in hand, will please the children in the house. To make it, pull a wire coat hanger into an oval frame, cover the frame with chicken wire sprayed with gold paint. Wire the clown to the center of the frame and decorate with artificial flowers and tinsel. Brightly colored aluminum foil flowers and tinsel rosettes were used on this one. Tie a bunch of balloons, that are really Christmas balls, to the clown's hand. Wire narrow streamers of ribbons in the colors of the flowers to the balloons. Wire a toy horn to the other hand. You may be sure that many of your guests will blow the horn to signal their arrival. This may be used on an outside door that has some protection or on the door of a family room for a note of fun and frolic in a Christmas decoration.

Fresh flower swags, Plate 15, make delightful decorations for apartment doors, or of course may be used outdoors in our warmer states. To make this one, wire bright red glads and blue-green Colorado spruce together in good line pattern. Insert the stems in a pillow of wet sphagnum moss covered with aluminum foil. Tie a group of green glass wine bottles together. Spray seven raffia glass covers with gold paint. Wire gilded sweetgum balls to the center of each for clappers. Wire these bells to the greens. Tie with a red velvet bow to complete this handsome door decoration.

Wreaths are all-time favorites as door and mantel decorations for Christmas. New ideas appear each year, and everyone enjoys making one as part of a decorating plan. Inspiration is everywhere; all we have to do is develop an awareness. This book is filled with ideas that may be copied, or for real satisfaction interpret them in your own way, perhaps using different materials or different colors to blend with your decorating plan and to express your taste and personality.

15. *Designed by:* Barbara S. Meisse
Photographer: Earl Roberts

16. How to make an evergreen wreath.
Photographer: Bryan Studio

There are many methods for making a wreath but one of the easiest, ways to produce a sturdy one is to use a florist's 4-wire frame as in Plate 16.

To make an evergreen wreath, stuff the center of the frame with wet moss and wrap in green aluminum foil. Tie securely with green string. Cut evergreens in uniform lengths, usually about 4 inches long. Cut the end of each twig to a sharp point. Thrust the greens into the wet moss for long keeping. First attach the outside row using the same greens all the way; then attach the inside row in like fashion. In the mid-section, use a variety of greens to relieve the monotony and add to the beauty of the wreath. In the case of a boxwood wreath, however, only boxwood is used. Cone

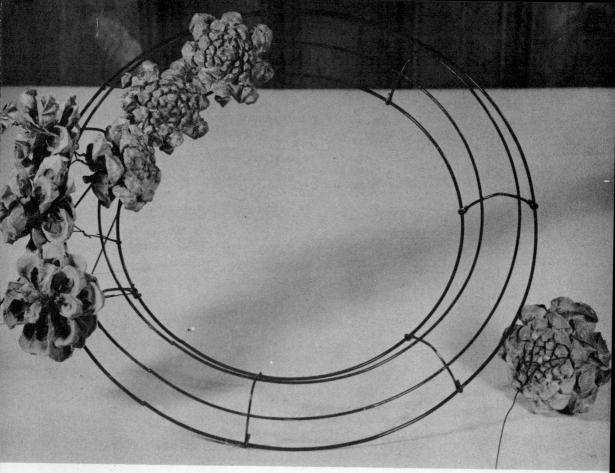

17. *Photographer:* Bryan Studio

wreaths or wreaths made of artificial material are made on the same 4-wire frame, without the moss stuffing. To make a cone wreath see Plate 17.

Choose cones of similar form and size for both the outer and inner rim of the wreath. Use larger cones on the outer rim. Wire each cone separately, leaving an end of wire to be attached to the frame, as shown in the plate. Wire the outside row on first. Place each cone against the outer rim, pull the wire on the cone over to the inside rim and wire it securely to the rim. Then wire the inside rim in similar fashion, pulling each wire on the cone over to the outside rim of the frame and wiring it securely to the frame. This sets up a cross tension that will hold the cones securely in place. Symmetry is desirable in a wreath, and this method will produce it. In the section between these two rows of cones, design and have fun. Use cut cones, seed pods, fruit, nuts, anything you desire to produce the effect you have in mind. Wire each unit separately and then wire all to the center wire of the frame.

The cone wreath in Plate 18 has a della Robbia center made of artificial fruit and pale yellow-green plastic leaves. Piñon (pine) cones from California form the outer rim and blue spruce cones are used on the inner rim. A chartreuse and blue-green two-toned satin bow completes the wreath.

18. *Designed by:* Author   *Photographer:* Bryan Studio

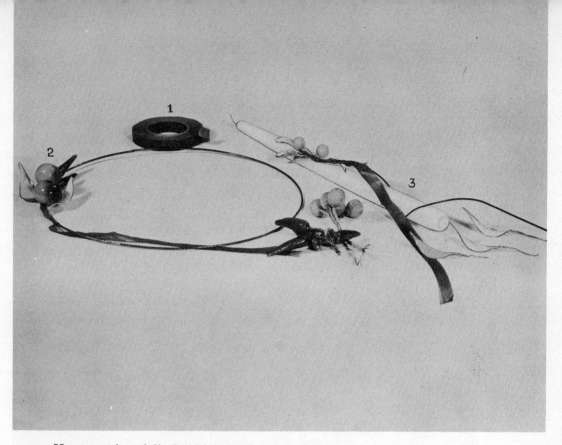

19. How to make a della Robbia wreath and candle spiral.
Photographer: Bryan Studio

To make the della Robbia center, use pliers and remove one wire ring from a 4-wire frame. Floratape the fruit and leaves to the ring as shown in Plate 19. Use floratape (1). Place the wired stem of the first piece of fruit on the ring (2) and tape it to the ring. Place the next piece a little lower and on the outside of the ring and tape it. Tape the next piece of fruit a little lower and on the inside of the ring. Group the colors and repeat the groupings of fruit for symmetry and unity of the wreath. Floratape the leaves in as you work. The fruit in Plate 18 is imported from Western Germany and the colors are muted tones of yellow, yellow-green and ochre accented by dark red bing cherries. To make the plastic leaves, see Plate 36 for directions. When the fruit and leaves are taped entirely around the single ring, wire the completed ring to the center of the cone wreath. Such a wreath may be used without further ornamentation as shown in the mantel decoration Plate 35, or it may be wired to the center ring of a 4-wire frame as shown in Plate 94. For an elaborate della Robbia wreath use fruit in place of the cones on the outer and inner wire frame.

20. *Designed by:* R. G. Schulke   *Photographer:* Bryan Studio

You can make the cone and nut wreath in Plate 20 without the tedious task of drilling the nuts so that they can be wired. Buy a thin gold mesh (tricotine, imported from France) from your florist or department store. Cut the mesh in 2 inch squares. Cover each nut with the tricotine, wrap a 22 gauge wire firmly around the mesh at the end of the nut, leaving a 3-inch piece of wire. Make groupings of three or four covered nuts and wire them to the center ring of a 4-wire cone wreath, as shown in Plate 19. Discarded sheer nylon hose, cut in 2 inch squares, may be substituted for tricotine. Nylon gives the nuts a groomed appearance, eliminates drilling them and of course costs nothing. This wreath was made of beautiful piñon (pine) cones from California. The backs of the cones are shown on the outer rim. They are reversed on the inner rim, showing their entire form, color and beauty. A two-toned chartreuse and blue-green satin bow was used to accent the sparkle of the tricotine and give importance to an otherwise very subtle color harmony.

21. *Designed by:* Author   *Photographer:* Bryan Studio

Plate 21 is another variation of the della Robbia wreath, done in the traditional Christmas colors, red and green. To duplicate it, use a green styrofoam wreath frame available at your florists. Cut variegated holly in 4-inch lengths. Insert these on the outside edge of the frame, covering it well. Wire pine cones separately and insert the ends of the wire into the inside edge of the frame. Make a della Robbia wreath using all red, or red and green artificial fruit, as shown in Plates 17, 18 and 19. When completed, wire it to the center of the styrofoam frame and tie with a bright red satin bow. After Christmas remove the fresh holly leaves and satin bow to store the wreath. Next year you need only add new leaves and a fresh satin bow to have a handsome wreath for your door or mantel.

22. *Designed by:* Mrs. Aubrey R. Holladay
Photographer: Bryan Studio

This beautiful wreath, Plate 22, is made of Texas tallow berries. Cut the twigs of berries in 3 inch lengths. Use a white styrofoam wreath frame. Insert the twigs into the frame, entirely covering it. Spray lightly with gold paint to give a pearly effect. Tie with a moss green satin bow. This is a wreath of elegance, appropriate for the Georgian door on which it was used.

Straw mats are used for the foundation of the door decorations shown in Plates 23 and 24. They can be used effectively in casual interiors. Very little material is required and this type of decoration is easy to make.

23. *Designed by:* Mrs. Edward F. Johnson
Photographer: Bryan Studio

24. *Designed by:* Mrs. Everett R. Combs
Photographer: Mr. Karl Hermann

In Plate 23 a straw mat is used as the frame of the semicircular fruit and cone wreath. It is simple and bold in design and makes an attractive door decoration. Spray three pine cones and a few glycerin-treated mahonia leaves with gold paint. Wire the cones and two bunches of artificial grapes to the mat; add the foliage and a satin bow to match the color of the grapes used. Fruit other than grapes may be used to give the color needed to carry out your particular color harmony.

A dramatic door arrangement suitable for Thanksgiving or Christmas is illustrated in Plate 24. It was designed for a walnut door and the materials were chosen in hues of light tan to golden brown. The mat is subordinate in this decoration. Compare it with Plate 23 and note how in both arrangements the principles of dominance and subordination are used to achieve the desired effect. To duplicate it, make a swag by wiring sorghum and wheat together in a diagonal line pattern. Wire two bunches of tawny green artificial grapes to this. For the center make a grouping of round artificial fruits and treated beech leaves and sew it to a piece of galvanized screen. Wire to mat.

25. *Designed by:* Mrs. Robert O. Evans
Photographer: Walter Bujak

For an inside door that harmonizes well with the nylon net curtain, see Plate 25. To make it, cut a circle of copper wire screen 27 inches in diameter. Cut blue-berried juniper and golden-tipped cedar into twigs about 4 inches long. Wire these together in groups of three and wire them to the outside rim of the circle of copper screening. Wire small cones and mixed nuts and combine them with Christmas balls in blue, rose and gold. Make into small clusters and wire to the center area of the copper screening. Fill in with bits of the evergreen. Wire gold, paper lace doilies to the background of the wreath. Make a bow with long streamers of narrow satin ribbons in blue, rose and gold to harmonize with the Christmas balls.

26. *Designed by:* Author
Photographer: Bryan Studio

27. How to make a lacelon wreath.
Photographer: Bryan Studio

Those who enjoy a feminine look in Christmas decorations will like the wreath, a bit on the fussy side, in Plate 26. Each medallion is made of three-quarters of a yard of gold and silver lacelon, available in variety stores or at your florists. A nylon draw thread is woven into this material, and you need only to pull it and wrap it around the stem of a glitter daisy to make rosettes.

Plate 27 shows how to make the wreath shown in Plate 26. Wrap a 4-wire frame with white floratape to conceal the wires (1). Cut three-quarters of a yard of gold lacelon for each rosette (2). Gather it into a rosette by pulling the nylon thread that you will discover running through one edge of the lacelon. Use gold and chartreuse glitter daisies made with chenille stems, also available in variety stores, for the center of each rosette (3). Wrap the nylon thread of the gathered lacelon around the stem of the daisy and tie it securely (4). Wire the rosettes to the outside and inside rims of the wreath frame. Make a bow with streamers of chartreuse satin ribbon and wire it to the frame for a contrast of textures. This wreath can be made in a variety of colors.

Mantels that Radiate the Spirit of Christmas

The mantel is often the focal point of a room and as such is the ideal place for an important decoration. Whether you plan to use a wreath, festoon, candles, Madonna or crèche, it should be well related in size, color, texture and spirit to your particular mantel. Somewhere in the decorating, it is well to tell the Christmas story and the mantel is a suitable place to do this.

The Christmas creche

The Christmas crèche never grows old; it combines the charm of antiquity with enduring youth. The crib and the Christmas tree seem to children a glimpse into fairyland, and Christmas has the magic power to make children of us all if we will let it. Children, young and old, gazing at the Christmas crèche, can stand in the manager of Bethlehem and seen the newborn babe with Mary and Joseph, the animals and the shepherds and wise men adoring Him. They find joy in it and wait from one Christmas to the next to renew this joy. The crèche comes to us through Christian art and tradition, from its beginning two thousand years ago in Bethlehem. St. Francis is often given credit for originating it, but it was twelve hundred years after the birth of Christ that St. Francis erected his crèche in Greccio. Many were made before his time, but records of them are brief, and St. Francis took old customs and traditions and gave them the stamp of his unique genius. Christmas for St. Francis was always a Feast of Hope and Light, of Peace and Joy and Brotherly Love as expressed in his crèche. He pointed the way for many to follow. Each year every country that celebrates Christmas portrays the birth of Christ in its own managers with its own familiar landscape as a background. The Madonnas and cribs pictured in this book come from many countries of the world and show a strong national influence.

The mantel in Plate 28 is in a studio-type living room requiring decorations large in scale. The antique Spanish crib is in the medieval color harmony of red, blue and gold, repeated in the fruit

28. *Designed by:* Author   *Photographer:* Walter Bujak

garland and medallions. The garland of artificial gold leaves, pink and blue-violet grapes is a modified version of the colors found in the crèche and in many of the famous medieval paintings of Madonnas.

To make the garland use 20-gauge spool wire. Cut a piece long enough to fit the mantel. Floratape the wired gold leaves and fruit to it in the same manner as the candle spirals are made in Plate 19.

To make the medallions for the candelabra, bend a 16-gauge wire into an S line. Floratape the gold leaves and artificial fruit to it in the same manner. Wire the finished medallion to the gold scroll and to the candelabra.

29. *Designed by:* Author   *Photographer:* Harry Wyatt Schulke

A simple theme well executed is better than an elaborate one poorly done. The 300-year-old handcarved Madonna and Child from Spain, in Plate 29, stands between two stylized trees of glycerin-treated foliage and artificial oranges.

Double treatment for a mantel usually satisfies the eye. The eye tries to find likes and so moves from one part of a composition to another to find them. By introducing a different material or form, in this case the Madonna and Child, the eye finds rest in the change. This method of designing sets up a tension that helps us to live in the design. The topiary trees are inspired by ornamental gardening. For the method of making them see Plate 30.

Any broad-leaved evergreen, such as boxwood, huckleberry, ilex or mahonia may be used to make these trees and it is best to treat the foliage with glycerin. To do this mix one pint of glycerin with one quart of water, put it in a large can or container. Split 2 inches of the end of each twig to be treated. Keep cut ends of branches in glycerin solution for about two weeks. The foliage will absorb the solution through the stems and preserve it for **years**. Treated huckleberry foliage is used in these 30-inch trees.

30. How to make a topiary tree.   *Photographer:* Bryan Studio

Plate 30 illustrates how to make the topiary trees shown in Plate 29. Cut 2-inch-thick green styrofoam pyramids, 20 inches tall and 7 inches wide. Cut the treated foliage in 6-inch lengths and insert in the styrofoam to make a pyramid. When the frame is completely covered, prune severely for a stylized effect. Decorate with artificial oranges. Place the completed trees on a large needle-point holder and set them in a pair of identical containers. Reproductions of old stove tops were used for these trees.

It is typically American that we have taken customs and artifacts from all the countries of this earth and made them into a pattern of Christmas that is unique. America is the melting pot of the world and these customs and objects of art that have been brought to us from many countries by homesick hearts are given new interpretations and new energy by many Americans who see the beauty in them. Our Christmas decorations are a potpourri of many beautiful customs of the world; no wonder it is such a colorful, joyous and beautiful season. Plates 28, 29, 31, 32, 33, 34, 35 are examples of this.

In Plate 31 the brightly colored contemporary figures, Mary and Joseph traveling to Bethlehem, are a delightful example of Mexican folk art. The sombreros make them really belong to Mexico and add to their charm. The brass roosters on the wall, completing the composition, were also made in Mexico.

To make these topiary trees use a 6-inch styrofoam ball. Cut glycerin-treated boxwood in 4-inch lengths and insert in the ball, completely covering it. Prune the boxwood for a stylized look. Cut a ½-inch wood dowel 24 inches long. Wrap with white floratape. Insert the dowel in the styrofoam ball and anchor it in a heavy needle-point holder at the base. Use inverted, white plastic refrigerator boxes, decorated with gold paper for the base of these trees and to cover the needle-point holders. Trim the trees with brightly colored fruits to repeat the colors in the figures.

31. *Designed by:* Author   *Photographer:* Bryan Studio

32. *Designed by:* Author   *Photographer:* Bryan Studio

Plant material available in northern United States was used in Plate 32, a mantel arrangement with a Victorian flavor. The green, brown and gold, handcarved Madonna, Our Lady of the Woods, determined the color harmony and the spirit of the arrangement. She is a copy of the Madonna found in the cathedral at Monterey, Mexico, and was placed in a Victorian glass bell, backed with golden-tipped cedar. White Christmas roses (helleborus niger) that can be found under the snow in many northern gardens were used in the focal area.

To make the wreath that encircles the glass bell, use a single wire frame. Wire garden hens and chickens (sempervivum) separately, spray them lightly with gold paint and floratape them to the wire frame. The moss green velvet ribbon tied to the brass candlesticks helps to unify the composition and adds to the Victorian spirit.

Sago palm or cycas, trimmed in symmetrical form and sprayed with gold paint, is combined with an illuminated wreath of gold lacelon and gold and green glitter daisies and artificial gold leaves to make the mantel arrangement in Plate 33. All these materials are available from your florist and in some variety stores. The wreath provides a second halo for the handcarved, natural wood, Mexican Madonna, and being illuminated produces an exciting effect. Note how the dragon wings of the brass candlesticks repeat the plant form of the sago palm, emphasizing the rhythm.

To make the wreath, use a 24 inch 4-wire frame. Place a string of 36 tiny white lights (Italian import) inside the frame. Wrap with gold lacelon. Wire both gold and green glitter daisies and artificial gold leaves to the outside ring of the wire frame. Insert the palm leaves in a block of styrofoam.

33. *Designed by:* Author   *Photographer:* Bryan Studio

34. *Designed by:* Author   *Photographer:* Bryan Studio

A Victorian clock case of wood, in natural finish, provides a niche for the handcarved Madonna—Our Lady of Health—in Plate 34. The original may be seen in the cathedral in Potzcuro, Mexico. This is a very simple but effective mantel treatment. The fruit and foliage in the candle spirals pick up the colors of the Madonna.

35. *Designed by:* Author   *Photographer:* Bryan Studio

To make the spirals see Plate 19 (3). Use an 18-gauge wire. Spiral it around the candle. Start 2 or 3 inches below wick to avoid danger of fire. Place the first piece of fruit on the wire and flora-tape it to the wire. Tape the next piece a little lower and to the left of the wire; the third piece a little lower and to the right of the wire. Tape the leaves in as you work. Group the colors to avoid spottiness. These spirals may also be made with small Christmas balls and tinsel rosettes in place of the leaves, in the fruit spirals.

The della Robbia wreaths and garlands, so popular in Christmas decorations, are inspired by the glazed terra cotta sculpture done by Luca della Robbia and his family in Italy in the 15th century. Examples of his simple, unaffected naturalism may be seen in many art museums. The dominant colors in his work are intense blue, yellow-green, purple and ivory. These are the colors

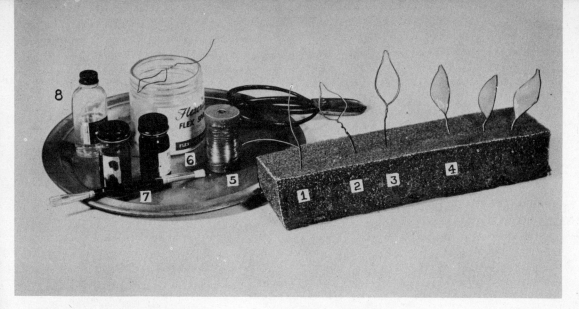

36. How to make plastic leaves.   *Photographer:* Bryan Studio

used in the wreath, Plate 35. Artificial fruit in beautiful muted colors is now imported from Western Germany. This provided the blue, yellow-green and purple. The problem was to find the ivory-white. The answer came in plastic leaves that harmonize perfectly with artificial fruit. The method of making them is simple, and you will find many uses for them. They can be painted any color that is needed for a specific color scheme. See Plate 36.

To make this della Robbia wreath, floratape these leaves and the artificial fruit to a single wire frame as shown in Plate 19. A modern, Italian porcelain Madonna was used in this mantel decoration. The color of the Madonna was dominantly ivory, shading into a warm brown that blended into the inverted, antique, bronze container that was used as a base. Ivory candles in copper and glass candlesticks, complete the composition.

Plate 36 illustrates the making of leaves. Cut 22 gauge copper wire into 7 inch lengths (1). Bend into a leaf form (2). Simple shapes work best. Dip the leaf form into clear, liquid plastic that can be purchased in a craft or hobby shop (6). Be sure to dip the leaf below the place where the wire is twisted, to insure the plastic spanning the leaf frame. Insert the leaf in a block of styrofoam to dry for several hours (3). When you purchase the plastic, buy a bottle of solvent to keep it thin and to extend its use (8), and bottles of paint made for this express purpose (7). For the ivory leaves, mix white paint with a bit of yellow-green and red. For a pleasing bronze green, mix dark green, yellow-green and a bit of red paint.

37. *Designed by:* Mrs. Aubrey R. Holladay
Photographer: Bryan Studio

When the transparent leaves are thoroughly dry, paint the back of the clear plastic the desired color and return it to the styrofoam block to dry again (4). You will note that since the leaves are transparent, the paint shows through on the front and leaves an attractive copper wire edge to the leaf. To give the leaves a third dimension, making them more convincing as leaf forms, twist the wire leaf form slightly after step two.

Carolers sing of Christmas and enjoy a never ending popularity in Christmas decorations. The merry group of English carolers in Plate 37 stand on a block of white styrofoam that has been edged in moss green satin ribbon to harmonize with the greens, the candles and the bows on the festoon. For a close-up of three of the figures and a step in making them see Plates 38 and 39.

The festoon over the mantel is easily made by wiring cones and seed pods to white pine roping that may be purchased from your florist.

These easily made English carolers can be used as a mantel decoration, on a Christmas Eve supper table, or on a door swag or wreath.

The Christmas tree in the background of Plate 38 is sorghum,

37. *Designed by:* Mrs. Aubrey R. Holladay   *Photographer:* Bryan Studio

steamed over a boiling teakettle to loosen the branches. Spray it with white paint and while still wet sprinkle with silver glitter. Note the interesting textural relationship of the sorghum and the rough tweeds and the striking contrast of textures against the glossy southern holly.

The human eye instinctively looks for likes, it finds rest in contrast. Note how true this is in this composition. Texture evokes a very definite emotional reaction. Can you recall how as a child you loved to touch fur, and the pleasant satisfaction that it brought to you? Use this knowledge to create the sensation you desire in your work.

To make the carolers (Plate 39) use a 4-inch piece of ¼-inch wood dowel for the body, and an oval wooden bead for the head. Paint the face and glue on bits of fur or yarn for the hair. Use 18-gauge wire for the arms and legs. Wrap the wire around the dowel as shown. Make clothes from small pieces of tweed, wool, felt and fur. They are exciting and fun to make.

39. How to make carolers.   *Photographer:* Bryan Studio

40. *Designed by:* Mrs. Richard E. Barnes
Photographer: Bryan Studio

Gold and moss green is the color harmony used for the mantel in Plate 40, in a charming Western Reserve farmhouse. The topiary trees are made by the same method as shown in Plate 30, but these materials are dried, chartreuse hydrangeas from the garden and glycerin-treated beech leaves sprayed with gold paint.

You will discover that paint sprayed on glycerin-treated foliage, does not chip or flake off as it is apt to do on fresh plant material. It is best to use treated foliage in Christmas decorations that are to last for several weeks.

The hydrangea is massed in the center of these trees and the gold beech leaves line the edge. They are tied with moss green ribbon bows and are in perfect harmony in color, texture and character with the moss green candles and the crystal and gold girandoles. This mantel expresses a spirit of generous hospitality —a holiday welcome to this Western Reserve farmhouse.

41. *Designed by:* Author   *Photographer:* Harry Wyatt Schulke

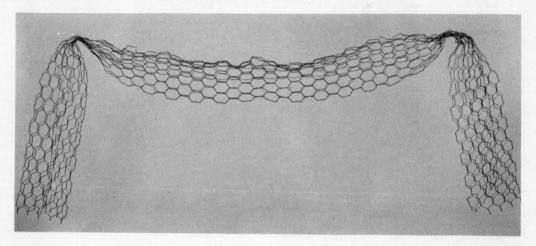

42. How to make a festoon   *Photographer:* Bryan Studio

Two large wooden, antique angels from Germany, Plate 41, holding a garland in their hands, were used above a mantel. The festoon is made of pin oak leaves treated with glycerin and sprayed gold, attached to a chicken wire frame that can be fashioned in shape and size to fit any mantel. (See Plate 42.)

The pink and blue-violet grapes and purple plums, symbolic of the peace and the promise of Christmas, develop a rich color harmony that is effective in many homes.

Plate 42 illustrates a step in making the chicken wire frame for the festoon, used in Plate 41. Cut the wire any desired length, roll it into a cylinder, flatten it and shape it into garland form. Twist the wire at the point where it is held in the hands of the angels. Wire the sprays of pin oak leaves separately and then wire them to the chicken wire frame, entirely covering it. Superimpose the fruit on the leaves and wire it to the frame.

43. *Designed by:* Mrs. Henry A. Sinderman
Photographer: Bryan Studio

Wreaths, symbolic of life eternal are all-time favorites for Christmas decoration. A handsome cone and seed pod wreath made on a 4-wire frame, edged with treated mahonia leaves, lightly sprayed with copper paint and tied with a copper metallic bow, makes a simple but satisfying decoration for this classic mantel, Plate 43. See Plate 17 for method of making this wreath.

Note how the length and placement of the candles and the line of the evergreens on the candlesticks point up the importance of the wreath, carrying the eye back to it. The function of design is to give order in space and here is an example of how pleasing it can be.

44. *Designed by:* Mrs. Paul McConnell
Photographer: Coleman Todd and Associates
Snow figures by: Mrs. Henry Paulsen

The Scotch pine and balsam fir used in the wreath and lantern decorations in Plate 44 beautifully express Christmas in northern Ohio. Daily fires on the hearth below do not affect this type of evergreen—something to consider when you select your greens for Christmas.

Mr. and Mrs. Winter, the white cotton figures, seem perfectly at home in this Early American Provincial setting. They are realistically decorated with doll hair, felt hats and miniature Christmas packages. They were made on wood and wire frames, similar to the one in Plate 39. The faces and bodies are made of cotton. The features, eyes, nose and mouth are paper. The hats are made of felt and put together with a great deal of imagination and skill. Because they were made by a person possessing creative ability, they seem to come alive and give a feeling of motion—note the man taking a step, and the coy but pleasant smile on the woman's face. Directions for making the evergreen wreath are given in Plate 16. White styrofoam snowballs decorate this one and white cotton batting provides the snow on which the figures stand.

45. *Designed by:* Mrs. Carl S. Graves
Photographer: John R. Walter

Plate 45 is another example of regional material, beautifully designed for a large mantel in the lounge of a hotel in northern Florida. Long leaf pine, palmetto branches cut to shape and sprayed with white paint and silver glitter, white gladiolus, coconuts and papayas painted in the same manner, all natives to Florida, are combined with the handmade candles to make a dramatic mantel decoration.

46. *Designed by:* Mrs. Raymond Russ Stoltz
Photographer: Arthur Heitzman

A delightful composition, Plate 46, suggests a frosty Christmas morning. It is arranged on the shelf of an 18th Century mantel that is painted a muted pistachio green, against gray living room walls.

The wreath of long needle pine is made on a 4-wire frame. See Plate 16 for method. It is sprayed with gold paint and then lightly sprayed with white paint and decorated with gold Christmas balls, then hung within the gilt mirror frame. The pine branches on the mantel shelf are painted in the same manner and the plastic deer are painted gold. A few gold Christmas balls are added to the greens to complete the composition.

The mantel decoration in Plate 47 combines a wreath of artificial material with topiary flower trees made of natural material. To make the wreath, paint a 36-inch 4-wire frame gold and cover it with gold net. Make eight small arrangements of straw flowers in various shades of pink, small velvet cerise primroses, red velvet forget-me-nots, moss green velvet leaves and gold and pink Christmas balls. Place these arrangements at every other cross section of the wreath frame and wire them to it. Wrap the wreath with 2-inch moss green velvet ribbon to garland the sections between the arrangements.

The picture of the Madonna of the Chair was purchased at a white elephant sale. The polychromed plaster frame was painted white and lightly sprayed with gold paint. The wooden angels that stand on the mantel shelf are antique gold.

The flower topiary trees at each end of the mantel shelf are made of dried Peegee hydrangeas from the garden. The color is a pale clear chartreuse green and pale pink to soft rose, the natural color of the dried flowers. To duplicate these trees, make them in the fall when the flowers are fresh and pliable. Hold one long stemmed Peegee hydrangea in your hand, add two or three more to it, holding them in place. Wind a string firmly around the stems. Add other flowers if needed until the tree is the size you desire. Any protruding flowers can be clipped out and small pieces added in hollows to give a symmetrical form. The stems become the trunk of the tree; warp these with brown floratape, stand in a needle-point holder and put in a warm, dry room until thoroughly dry. When you are ready to use the trees, put them in small brass pots. Spray with clear lacquer and sprinkle with gold glitter. Decorate with small, pink Christmas balls.

47. *Designed by:* Mrs. William Siemon
Photographer: William E. Buvinger

48. *Designed by:* Mrs. Robert Mannfeld
Photographer: Kenny Strattman

In Plate 48, a handsome mantel decoration is made entirely of garden material, white pine, magnolia leaves and Gilbert's celosia in values of red, blending from pale grayed rose to darkest maroon. It was made on a 4-wire frame. (See Plate 16 for method.) Dried celosia keeps for years. The large brass candlesticks are decorated with the same plant material and tied with gold satin bows to repeat the color harmony. (To dry celosia, cut the flowers when they are at their best and before seeds form. Hang upside down in a dry dark room to retain their color.)

49. *Designed by:* Mrs. Donald L. Millham
Photographer: Bryan Studio

Plate 49. Seven pewter mugs in classic design are used as a unit in this distinctive mantel arrangement. Pine branches from the arranger's garden are used with mellow toned piñon (pine) cones. These are carefully arranged in the three largest mugs. The pine is then cadenced from the fourth through the seventh mug, giving the necessary long line to balance the arrangement. Treated beech leaves and the cones on the pine branches give a contrast of form and color to the arrangement and harmonize beautifully with the fireplace and the paneled wall.

Candles, the little lights of Christmas, give charm and distinction to the three mantel decorations in Plates 50, 51 and 52. It is said the custom of using candles for Christmas came to us by way of Ireland and that the Irish people gave them the name of *Little Lights of Christmas.* Whether or not this is true, it is a pleasant thought. Today candles are used in every country that celebrates Christmas, in homes and churches alike. Their warm friendly flames express the brightness of this happy season and warm the hearts of all who see them.

50. *Designed by:* Mrs. Edward F. Johnson
Photographer: Bryan Studio

The horizontal swag of evergreens, cones and seed pods is dramatized by the vertical placement of the candle groupings in Plate 50. The placement of the gold stars, slightly off balance, gives life to the design. The large candle and cone wreath on the table below repeat the theme and the glow of the candles gives a feeling of warmth to the room when there is no fire on the hearth.

To duplicate this mantel, place the candles in a large block of styrofoam. Make the cone and evergreen swag on a chicken wire frame. Place it on the mantel shelf to cover the styrofoam base. Rolled gold metallic ribbon repeats the color of the stars. See Plate 17 for directions to make cone wreath.

In Plate 51, an interesting grouping for a mantel, we have three units well related to each other and to the space they occupy. Note how the top point of the star leads your eye away from the straight row of candles and down again to the pyramid, nut and teasel tree.

The lemons and pears used on the candelabra were dipped in paraffin. This gives the fruit a frosty appearance and insures its lasting through the holidays. Seed pods from the garden, a gilded artichoke and tricotine-covered nuts are wired to the candelabra. A few teasels are added to the gold paper star.

The nut and teasel tree is made on a pyramid of styrofoam as shown in Plate 30. Cover walnuts with gold tricotine, wrapping a wire around them. Make rosettes of the gold tricotine and wire these also. Insert the nuts, teasels and rosettes into the styrofoam pyramid, covering it completely. Place on a gold paper star for a composition that has variety within unity.

51. *Designed by:* Emily Stuebing   *Photographer:* Paul Engler

52. *Designed by:* Mrs. John Downing
Photographer: Richard Squires

Copper lustre was the color chosen for the contemporary mantel
treatment in Plate 52. Aspidistra and sea grape leaves with lotus
pods were sprayed with copper paint. The circular copper tray expresses the infinity of the Christmas wreath. The stylized copper
angels from Mexico, holding lighted candles, climax the composition.

A great variety of decorations for mantels are presented in this
book. Each one is designed for the type of home in which it was
used. The proper scale and colors were a consideration in each
composition; but more important is the fact that each mantel radiates the spirit of Christmas.

Christmas Trees that Grow with Imagination

Christmas trees bring fairyland into our homes and make children of us all. Martin Luther is believed to have had the first Christmas tree. Legend tells how on Christmas Eve, after walking under the starlit sky, he hurried home and set up, for his own children, a tree with many candles, symbol of the starry heavens that had sent forth the little Lord Jesus on that first Christmas Eve.

The large evergreen tree holds the place of honor in Christmas decorations; but in many homes there is room for one or more unique trees. In the entrance hall, on the mantel, in bedrooms or on dining tables they add beauty and interest to the holiday décor.

The large evergreen tree is usually the traditional tree, with the same decorations, used in the same way, year after year. This can be beautiful. Let good taste govern the decoration of your tree. Group your colors and choose a dominant color harmony. Avoid using many colored lights, and in their place try all white lights or omit the annoying strings entirely and floodlight the tree from below for a dramatic effect. The most fun of all comes in dreaming up a unique tree for your home each year. You will discover that family and friends look forward each year to the new and exciting tree that you will produce.

Trees and wreaths for the Advent season, the four Sundays before Christmas, have long been used in Europe, and are now becoming popular in the United States. An Advent tree is usually presented to each child in a family. He loves it for it helps to mark with pleasure the time between Thanksgiving and Christmas, which for a child is an eternity. A candle for each of the four Sundays in Advent is placed on the tree, and every Sunday night at supper time the proper candle is lighted with a little ceremony of the family's own choosing. Sometimes a fifth candle is added for Christmas Eve. If you plan to make one for your child or grandchild, just let your imagination take over and know that he will enjoy the charm of this little tree.

53. *Designed by:* Author   *Photographer:* Bryan Studio

The Advent tree in Plate 53 will delight children and add beauty to the pre-Christmas season. It is made of wood and could be duplicated by anyone who is handy with tools. This one is 27 inches high and 20 inches wide. Nail a square board to the base of the triangle to hold the crèche. Make the candle holders from strips of tin cut from cans and nail them to the sides of the triangle. Spray the tree with gold paint. The candles were decorated with wheat to suggest the humble birth of Christ. The ribbons that tie the wheat repeat the colors in the figures—moss green,

54. *Designed by:* Author   *Photographer:* Harry Wyatt Schulke
Courtesy of Flower Grower Home Garden Magazine

chartreuse and pink. A similar crèche may be found in variety stores.

In Plate 54, the Advent tree is Swedish in origin and is used that last long week before Christmas. Seven birds in a box accompany the tree when it is given to a child, and he places one on the tree each day. The top one is reserved for Christmas Eve. It is obvious that the children under this tree are being very good for Christmas, and even the cat, paying absolutely no attention to the birds, has been awarded a gold star halo for good behavior.

55. *Designed by:* Mrs. Glenn DeHoff
Photographer: Paul M. Clapper

The contemporary version of the Advent tree in Plate 55 is made of wrought iron and expandable metal and is available in gift shops. Decorate it in any fashion you desire. Here the white candles provide a striking contrast with the black tree. Scotch pine, pink pepper berries and pink Christmas balls with tinsel puffs make up this arrangement. It will last all through the Advent season.

Small artificial trees available in variety stores and a box of 24 small ornaments for the child to place on the tree each day of the Advent season will be enjoyed by many children. By counting the number of ornaments left in the box the child can answer for himself the often asked question, "How many more days until Christmas?"

56. *Designed by:* Author   *Photographer:* Walter Bujak

The color, movement, music and many toys on the tree in Plate 56 will delight children from 6 to 60. It is a modern interpretation of an early German tree called a Weinachtspyramide. The heat arising from the lighted candles causes the windmill to turn and the shelves to revolve. The pyramid is 4 feet high and 27 inches wide. It is mounted on a large Regina music box and both are painted Swedish red. This would have to be made by someone skillful with tools.

Garlands of green leaves, red artificial fruit and mushrooms, decorate the supporting rods. On the top shelf is an angel band. The second shelf is a garden. A gay Christmas parade marches around the bottom shelf. The figures are small wooden toys, imported from many countries. They are fun to collect when you travel, but may also be found in our shops at home. Each year new toys are added and family and friends look forward to seeing

57. *Designed by:* Author   *Photographer:* Walter Bujak

them. Plate 57 is a close-up showing detail of the garden and the
Christmas parade. This tree is used on a chest in the dining room
of an American Provincial home. Brightly colored candelabra and
brass roosters on the wall complete the composition. Just light the
candles and start the music box. As the shelves revolve, a merry
little tinkling tune sets the stage for a happy holiday meal.

58. *Designed by:* Author   *Photographer:* Harry W. Schulke
Courtesy of Flower Grower Home Garden Magazine

In Plate 58 is the Italian version of the pyramid tree, called a
ceppo. A ceppo is given to each child of an Italian family and
friends and parents secretly place gifts on it. Our modern interpretation of such a tree is painted gold and decorated with artificial fruit and a straw crèche. This one is 48 inches high and the
sides of the bottom triangle shelf measure 20 inches. The next
shelf measures 17 inches, the one above it 12 inches and the top
one 8 inches. Drill holes in the shelves and run 1½ inch dowels
through the holes. Drill additional holes for the candles. Following the tradition of the ceppo, place a cone or a puppet on the top
shelf, and fruit and candy on the upper shelves. The lowest shelf
is usually reserved for the Christmas crèche.

59. *Designed by:* Author   *Photographer:* Harry W. Schulke
Courtesy of Flower Grower Home Garden Magazine

Use this same frame to make a toy tree for your child. He will
be pleased if you use some of his old toys as well as new ones.
Plate 59 shows a close-up of a shelf of a toy tree.

Almost every traveler who goes to Mexico brings back some of
the charming, colorful figures of woven palm that the Indians
make for toys for their own children. The tree in Plate 60, made of
sago palm painted bright pink and decorated with strings of ma-
genta palm beads, boasts a band of mariachis and a group of farm-
ers working in the pineapple fields.

60. *Designed and photographed by:* Harry Wyatt Schulke

The frame of the tree could be used for many types of trees. To make it, cut a piece of hardware cloth in a semicircle to make a cone 50 inches high and 30 inches wide at the base. Wire it together with 20 gauge spool wire. Use a broomstick for the trunk and wire it to the top of the cone, and place it in a container. Fill the container with plaster of Paris to hold the tree securely. An inverted stove top, painted pink, was used for this tree.

Treated sago palm or cycas can be purchased from your florist. Paint it pink and insert it in the holes of the wire mesh cone and wire them to it. Put on a row at a time, covering the frame. Wire the palm toys to the tree and decorate with the palm beads.

61. Designed by Mexico's oldest pinata maker.
Photographer: Bryan Studio

The colorful paper pinatas (Plate 61) that can be purchased in
every Mexican market are really the Mexican Christmas tree. The
American Christmas tree is beginning to find favor in Mexico, but
the traveler to this enchanting country finds himself hoping that it
will never supplant the pinata. These ceramic and paper creations

are usually made in the form of animals or birds. The body is a hollow ceramic jug that is filled with goodies and small gifts for the children. Colorful papers cover the body and form the rest of the animal. This one is encircled with a hoop of gaily colored paper flowers. Mexican children look forward with pleasure to the breaking of the pinata at every fiesta in which they have a part. It is fastened to a rope that is suspended between two poles on a pulley. The rope is slack and the adult who manages it, can raise or lower it, depending on the height of the child who is trying to break it. Each child in turn is blindfolded, provided with a heavy stick and given three or more chances to break the pinata. When some lucky child accomplishes this, the goodies fall to the ground and there is a wild scramble in which each child tries to get as many gifts as possible. It is fun just to watch and one can easily imagine how much fun it is for the children who look forward with anticipation and pleasure from one pinata to the next.

The bird Christmas tree in Plate 62 is really a collector's item. It would delight children and adult bird watchers alike, for 35 different kinds of birds rest on its branches. They are made of colored felt and are the creation of Delia Franklin Castor of Ponca City, Oklahoma, who is famous for her birds. The tree was made by nailing an interesting branch to a slice of wood with bark left on it. It was painted white and the birds were wired to the branch.

62. *Designed by:* Mrs. Delia Castor
Photographer: Edward A. Bourbon

63. *Designed by:* Mrs. Daniel Bishop
Photographer: Donald J. Porter

Individually designed trees for dining tables and coffee tables add charm to any home. The unusual tree in Plate 63 originated in Florida and is made by braiding the stems of the bloom of the Cocos Plumosa palm. It is called a Palm Epergne tree. The jewel-like, hand decorated eggs are a perfect decoration for this epergne. Eggs were one of the first decorations used on Christmas trees and many beautifully decorated ones have been preserved. It may well be that they served as the inspiration for the delicate, jewel-like eggs that are being made today.

To make these ornamental eggs: 1. Draw an oval line from top to bottom of an uncooked egg. 2. Pierce a hole in top center of this line with a large darning needle. 3. With a sharp manicure scissors, starting from this hole, commence cutting around the line. Remove contents of egg, rinse shell in cool water. Dry with a pad of cotton. 4. Dip the shell in egg dye or cake coloring if you prefer a colored ornament. 5. Apply one or two coats of clear nail polish to inside and outside of egg to help preserve it. 6. String a double gold cord through a hole in the top of egg and knot the ends inside the egg, making a loop to hang it. 7. Glue a paper lace doily or Christmas scrapbook picture, or a tiny scene taken from a

64. *Designed by:* Author   *Photographer:* Bryan Studio

Christmas card, to the inside of the egg. 8. With tweezers, glue tiny gold stars around the doily. Use a toothpick to paste gold braid onto the rim of the opening. 9. Pour a few drops of melted paraffin into the bottom of egg, insert figure into wax to secure it. 10. Glue a tiny ropelike piece of cotton to rim of opening of egg. When dry put a few drops of glue on top of cotton and apply plastic pearls, available by the yard. Other jewels may be added. Make them as simple or as elaborate as you wish.

A Partridge-In-A-Pear Tree, Plate 64, is the theme of this modern version of the Swedish cookie tree. The Scandanavians use them on holiday smorgasbords to hold wreath cookies. We may use them on the Christmas breakfast table to hold powdered doughnuts, or on the coffee table filled with pretzels to accompany holiday cheer.

To duplicate this popular little tree, cut two pieces of ¾ inch wood doweling into 18-inch lengths. Cut one piece 24 inches long. Cut a 1 x 2 inch board 18 inches long and drill three ¾ inch holes in it. The center hole should be drilled through the board and the longest dowel inserted through it. Glue the outside dowels in the other two holes. Cut a 2 x 4 inch board 4 inches long, drill a ¾ inch hole in the center of it and insert the middle dowel in it. Paint Swedish red, black or any desired color. This tree was painted Swedish red and trimmed with artificial yellow-green pears, and dark green plastic leaves. For directions to make these leaves see Plate 36.

65. *Designed by:* Mrs. Howard M. Oberlin
Photographer: Howard M. Oberlin

The pine cone tree in Plate 65 was used on a table in a pine-paneled family room. It will keep for years and may be used natural, as shown here, or decorated with Christmas balls to carry out a desired color scheme. To make the tree, study Plates 66, 67 and 68. Cut a disc of masonite 24 inches in diameter. For the trunk, cut a piece of natural pine 6 inches long. Cut a 2 x 4 inch board 4 inches long. Screw the disc and the 4 x 4 inch board to the trunk of the tree. Cut ½ inch dowel 18 inches long and insert in a block of wood, screwed to the top of the disc as in Plate 66.

Wire white pine cones separately, leaving about 8 inches of spool wire at the end. Wrap these wires around the center dowel as shown in Plate 67. When this row is completed, wire additional rows on top of it. Start to build from the bottom and shorten the wires for each row to give a tapered effect. Wire a single pine cone upright, for the top, Plate 68.

To remove resin from cones, soak them in dry cleaning fluid for one hour, then brush them with linseed oil and allow them to dry for several days before using them. This brings out the color of the cones and helps to preserve them.

66. How to make a cone tree (1)
Designed by: Mrs. Howard Oberlin
Photographer: Howard M. Oberlin

67. How to make a cone tree (2)
Designed by: Mrs. Howard Oberlin
Photographer: Howard M. Oberlin

68. How to make a cone tree (3)
Designed by: Mrs. Howard Oberlin
Photographer: Howard M. Oberlin

Holiday Table Toppers

The dining table around which the family and friends gather becomes the center of holiday celebration. The color and decoration can set the mood. Lightheartedly let yourself go as you plan your Christmas tables. With the accent on casual living today, we may free ourselves from stuffy dogma and embark on an adventure of fun and creative satisfaction as we set our Christmas tables. To achieve the distinction looked for in any well planned decorative scheme, the decoration must harmonize with everything used on the table and with the room itself.

The color scheme for the table should tie in with the over-all color plan of the Christmas decorations for the entire house. Let

69. *Designed by:* Author   *Photographer:* Bryan Studio

color set the holiday mood that you desire. The entire decoration, including the candles and accessories, should occupy not more than one-third of the table surface. Avoid adding things that clutter up the table. Plan the decoration in relationship to the total effect. Distinction, originality, tasteful selection plus skillful execution will produce Christmas tables of gaiety and beauty. Table centerpieces that may be made well in advance to keep all through the Christmas holidays are the wise choice for busy women who like to spend as much time as possible with families and friends. Plates 69, 71, 72, and 73 are examples of arrangements that last.

Plate 69 is a Christmas ornament compote tree. It may be made in any desired color harmony. Use your own cherished Christmas treasures to give it personal charm. Choose a compote for the container to give it tree form. The variations in this decoration are limitless. You may choose a silver, pewter, brass, copper, glass or pottery compote. The materials may be all artificial or a combination of artificial and treated material. Or you may choose to add fresh foliage that keeps well, when the holiday arrives, as was done in the arrangement pictured. This is an easily made and most effective table decoration.

How to make the Christmas ball compote tree Plate 70. Cut a pyramid of green styrofoam that is 2 inches thick, the desired height to be in good scale with your compote. This one is 11 inches high and 5 inches wide at the base. Glue this to the compote or anchor it in a large needle-point holder. Gold glass rays

70. *Designed by:* Author   *Photographer:* Bryan Studio

71. *Designed by:* Mrs. Robert O. Evans
Photographer: Walter Bujak

purchased in a variety store were stuck into the styrofoam first
to define the shape and size of the decoration. Use any similar
material that you desire. Now from your boxes of Christmas ornaments select a variety of balls, angels, Santas, et cetera, well related in size and color. Balls made on wire stems are best for
this purpose. Insert them in the styrofoam, keeping the pyramid
form, using the smaller ones at the top and the larger ones at the
base. Complete the design on all sides. Last insert the foliage to
unify the design and to hide the styrofoam. Artificial leaves,
treated or natural foliage may be used. A brass Victorian card tray
was used for this tree. Gold, green and chartreuse ornaments were
chosen with bunches of bright fruit for accent. Fresh variegated
holly that lasted through the holidays was added as Christmas
drew near.

In Plate 71 is a compote table tree made of fresh material. The
method of making is the same as Plate 70. A milk glass compote
and candlesticks were chosen for this well designed decoration in
symmetrical balance. Insert wooden picks into the bright red lady
apples and thrust these into the styrofoam. Use euonymus foliage
from the garden to develop the traditional Christmas color harmony of red, green and white.

72. *Designed by:* Mrs. Everett R. Combs
Photographer: Mr. Karl Hermann

A table arrangement made entirely of artificial material is shown in Plate 72. The wreath is made on a single wire frame. Moss green velvet rose leaves and red artificial roses are florataped to it. The method of making this wreath is the same used in Plate 19. Place a round mirror inside the wreath. Make a nosegay arrangement of green velvet leaves, red roses and silver glitter daisies and insert in a needle-point holder. Cover with an inverted brandy snifter. Place a white twisted, glittered candle on top of the snifter and decorate with a spiral of red roses and velvet leaves, made by florataping them to a 16-gauge wire as shown in Plate 19 (3).

83

73. *Designed by:* Mrs. Howard Oberlin
Photographer: Howard M. Oberlin

Decorated fresh pineapples make most effective table arrangements. They may be used singly or in pairs, placed flat on the table or used on compotes. Spray the pineapple with paint to match your color scheme. The one in Plate 73 was sprayed with gold paint. String gold Christmas beads onto pearl headed corsage pins and insert in the center of every eye of the pineapple. Insert a sheaf of gold beaded rays in the top of the pineapple. Place on cedar twigs sprayed gold. The pineapple keeps well and since it has long been a symbol of hospitality and generosity, it makes a most appropriate Christmas table decoration.

The inspiration for this unusual table decoration, Plate 74, was the star candle, a replica of those used at the Moravian College in Bethlehem, Pennsylvania, at Christmas. The fruit and leaves were dipped in wax to harmonize with the candle. The method used for waxing the fruit is a very old Mexican technique.

Melt two pounds of paraffin in a large fruit juice can placed in a pan of water to prevent fire. When completely melted add one white crayon or one tablespoon of white lead paint. This gives the fruit an opaque glow. Have the fruit at room temperature and dip in the melted wax. It will keep in perfect condition through the holidays. This arrangement was placed on a star-shaped mat of green felt, used over a cloth of green nylon net.

74. *Designed by:* Mrs. William G. Wheeler
Photographer: Ray Jacobs

The covering for the Christmas table in Plate 75 is green nylon net over a Christmas green linen cloth. Green and gold rickrack braid cover the seams.

The abstract trees are cut from blue-green and yellow-green felt and stitched with gold thread. Gold snowflake sequins glued to the net help to unify the design. The fresh gardenia corsages in the compote tree repeat the white of the bone china dishes and may be used as favors for the guests. This tree is placed on a smoke glass compote of modern design. Use any broad-leaved evergreen foliage for the tree. Pittosporum is used here. Cut it in 3- to 5-inch lengths and thrust each twig into a pyramid of green styrofoam as in Plate 30. The tailored gardenia corsages may be purchased from your florist. Insert the wired stems into the styrofoam tree. Gardenias on the candlesticks carry the decoration down the table to complete the symmetrical balance. Carnations in any desired color could be made into boutonnieres and used in place of the gardenias.

75. *Designed by:* Author   *Photographer:* Bryan Studio

A handsome red and white antique damask cloth provides the perfect background for a Christmas table in an Early American house (Plate 76). White stoneware dishes and compote are right in color, texture and character for everything used on the table and for the room itself. The flowers are bright red geraniums used with ivy from the garden; a simple arrangement which is best for the patterned cloth. Get out your cherished heirlooms and use them at Christmas to establish never-to-be-forgotten traditions in your family.

76. *Designed by:* Mrs. Charles Spargrove
Photographer: Bryan Studio

The horizontal line design in Plate 77 lends distinction to an unusual Christmas luncheon table. The color inspiration came from snow and ice. An aqua-blue linen cloth was used with white china dishes patterned a deeper aqua. White carnations and pine branches used with simplicity and restraint, in the Japanese manner, produce an arrangement of natural beauty.

77. *Designed by:* Mrs. Donald L. Millham
Photographer: Bryan Studio

The della Robbia theme provides the inspiration for the table arrangement in Plate 78. Fresh fruits, in the medieval Christmas colors of red, blue and gold are used in the lower section of the alabaster epergne. Flowers in lighter values of the same hues are used in the top section. This type of table decoration is appropriate for a house using the della Robbia theme for the over-all Christmas decoration plan.

78. *Designed by:* Author   *Photographer:* Bryan Studio

79. *Designed by:* Robert Putt   *Photographer:* Bryan Studio

The table in Plate 79 presents a variation on the della Robbia theme, designed with great strength and beauty. The fibre cloth and the wood plates and serving pieces are made in the Hawaiian Islands. The tumblers are wood textured pottery. The arrangement is made of Wenlandi foliage, Orange Delight roses, orange mangoes, pineapple, bananas and dried poinciana pods. The star-shaped pieces on the table are large sea urchins from Hawaii. This table is set for a Christmas Eve supper in a bachelor apartment.

80. *Designed by:* Author   *Photographer:* Bryan Studio

   The Christmas punch table, Plate 80, makes use of a sort of
treasure that many people like to collect. The cloth is a hand-
woven red, and white coverlid. The colorful candelabra came
from Mexico as did the straw stars under them and under the
punch bowl. Four groupings of brightly colored artificial fruit
were made, repeating the colors in the cloth and candelabra.
These were wired to the straw star. It is gay and colorful and sets
the stage for a merry holiday.

Christmas Is for Children

A large part of the joy of Christmas is the fun of preparing your home for the occasion. It is wise to start early and let every member of the family help. Everyone loves the excitement of Christmas, the fun of decorating a tree and making ornaments for it, of making a wreath or swag for the door, or a gay decoration for the mantel or dining table.

Christmas preparations are an important part of every child's life. Working together in preparing for the day, the family discovers the true meaning of Christmas. If you have small children in your family, plan the decorations for them. Use their toys on the door swag, on the mantel and dining table. Use the decorations they make even if they are not quite perfect. These things delight a child and make him feel that he has a real part in the fun.

Many garden clubs are doing excellent and rewarding work with children, teaching them the joy of gardening and flower arranging. Carrying this project further, the Lakes and Hills Garden Club of Mt. Dora, Florida, under the direction of Mrs. Carl Sumner Graves, has conducted Christmas Workshops in their elementary schools for several years. They have been so successful that they have the support and interest of the school authorities and art teachers. The plan for these workshops is given here, with the thought that other garden clubs might find satisfaction in bringing this same pleasure to children in their communities. Quotation from the Florida Gardener:

> "The children who attended these workshops were just emerging from their belief in Santa Claus and so were invited to *Follow Santa's Footsteps,* and share in the spirit of Christmas by making others happy, just as Santa does. Giant red footsteps marked the path to the room where the workshop was held. Each room had the privilege of choosing a boy to act as Santa's elf. On the stage was a bare Christmas tree. The instructor made each gift and ornament, carefully showing the children just how it was done, and as they were finished the elves trimmed the tree.

82. Children's Christmas Workshop
Courtesy of Mrs. Carl S. Graves
Photographer: Bryan Studio

81. Children's Christmas Workshop
Courtesy of Mrs. Carl S. Graves
Photographer: Bryan Studio

There was a gift for each member of the family—for the neighbor's new baby, for a classmate who was home with mumps or measles, and a welcome for the new family in the neighborhood.

Each child received a mimeographed booklet containing instructions, pictures and patterns for making the gifts and ornaments. Each year additional pages with new ideas are added. The garden club furnishes all materials, and the teachers and garden club members follow up the program with workshops, supervising the children as they make these gifts for their own family and friends."

The ideas are not all original, but the old ideas were presented in an original manner. Plates 81 and 82 show some of the things that the children made. If your garden club "Follows the Footsteps" of this excellent plan to develop the true spirit of Christmas in children, you will think of many more ideas and add to them each year.

A Key Chain for Daddy. The transparent plastic tags may be purchased in a variety store for five cents each. The children brought small snapshot pictures of themselves and pasted them on the cardboard. Below the picture they wrote, *I Love You Daddy* and on the back of the card, *Drive Carefully.*

A Charm Bracelet for Mother. The chain and metal rimmed cardboard tags were purchased in a variety store. Snapshot pictures of each member of the family including the pets, were pasted to the tags. Tiny Christmas bells were wired to the chain to make it sing of Christmas.

A Friendship Bracelet. Use a charm bracelet or purchase a chain in a variety store. Paste a snapshot of yourself on a metal rimmed cardboard tag and wire it to the bracelet. Exchange snapshots with your friends and write your message on the back of the tag.

A Christmas Pussycat ornament for the tree. Cut whiskers and eyebrows from red, Christmas, wrapping paper and pin to a styrofoam ball with sequins to make the eyes, nose and mouth. Cut red pipe cleaners into four inch lengths and bend and thrust into the ball for ears. Add an extra pipe cleaner to hang the ornament to the tree.

A Gift Carry Case. Paint a soft drink carton a bright, pretty color. Paste cut-outs from Christmas cards on for decoration. Fill the carton with crayons, paper, scissors and paste for a shut-in friend or fill it with small toys for a little brother or sister. Decorate one in colors to match your livingroom and give it to Mother to display her Christmas cards.

Fortunate is the home that has children in it at Christmas. Everyone can invite children in to enjoy a candy tree from which they may choose a cane or sucker. Many types of trees can be transformed into candy trees. The welded espaliered tree in Plate 83, made especially for children, could provide the frame for many types of Christmas trees. It would be especially suitable for a della Robbia tree.

This tree is 60 inches tall and 42 inches wide at the base and was made by welding one inch iron rods together in this form. Wire the beautiful French candy suckers that are found in the shops at Christmas, and wrapped peppermint candies, to the frame. Make bows of gaily colored ribbons and wire them to the

83. *Designed by:* Author
Photographer: Bryan Studio

84. *Designed by:* Author
Photographer: Bryan Studio

frame to unify the design. Hang candy canes over the branches and renew the supply as needed. The jolly, felt Santa was wired to the tree last of all and he winks a merry welcome to everyone.

Cookie trees of all kinds delight children at Christmas. The cookies in Plate 84 were made from antique Dutch molds and were glued to a plywood obelisk, painted Swedish red. To make the obelisk, cut 3 triangles of plywood 48 inches tall and 15 inches at the base. Mitre the edges and nail together and paint any desired color.

Eyecatchers and Conversation Pieces

Unique ideas in suspended decorations, mobiles, kissing balls and Christmas card displays appear each year. They are a minor part of the over-all plan but when cleverly done often steal the Christmas decorating show. They are fun to dream up and ideas may come at any time of the year and in the most unexpected places.

Each year millions of Christmas cards arrive in homes everywhere. Many of them are beautiful and we like to share them with others. If we string them out on doors, windows or mantels we usually arrive at a cluttered effect that detracts from the beauty of the decorating plan. A simple and enjoyable way of sharing cards is to decorate a basket of suitable size, place it on a convenient table, put all your cards in it and invite your friends to look at them if they care to. This sometimes serves as an "ice breaker" in the mixed groups that have a way of gathering during the holidays.

85. *Designed by:* Author   *Photographer:* Bryan Studio

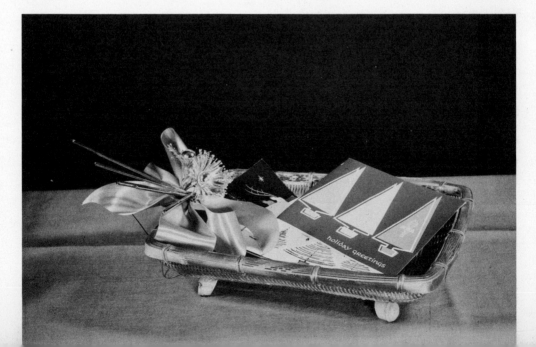

86. *Designed by:* Mrs. Donald L. Millham
Photographer: Bryan Studio

The Christmas card basket, Plate 85, was sprayed with gold paint and decorated with a chartreuse bow, gold glass rays and gold glitter daisies that were purchased in a variety store.

The gay Christmas card holder in Plate 86 presents a situation in reverse. The two little birds seem to register amazement over the fact that their home has been usurped for the holidays.

The bird cage was purchased at a white elephant sale. It was sprayed with copper paint, as were the cones and pine that are tied to the swing with a pink metallic bow.

The cards are piled high in the cage in gay confusion, but it is controlled and confined. What fun it would be to open the door and take out the cards for inspection.

Mobiles and suspended decorations have enjoyed great popularity in the past several years. Many of them are complicated and difficult to make, but here in Plate 87 is one so simple that anyone can make it. The body is a lightweight wire coat hanger, bent into shape and soldered or welded at two joints. The red cotton flannel boots and mittens and the materials used in Santa's head were found in a variety store.

To make the Santa head, use a 3-inch styrofoam snowball. Cut a half circle of red felt for the cap. Roll into a cone and glue it to the head. Edge it with white chenille wire. Glue this along the edge of the cap. Thrust the ends of the wire into the ball at the back to hold it securely. Cut 4 pieces of chenille wire 3 inches long and roll them up snail fashion for the mustache and beard. Thrust the ends of the wire into the ball just below the nose for the mustache and the beard below that. Use a small pink Christmas ball for the nose. Pin blue star sequins on for eyes. Sew sprigs of mistletoe to the mittens. Hang the *Little Santa Who Wasn't There* in a doorway, using black nylon thread. This mobile kiss catcher adds fun to the holidays and is guaranteed to bring results.

87. *Designed by:* Author
Photographer: Bryan Studio

Here in Plate 88 is a new version of the mistletoe ball. Soak white yarn in wallpaper paste. Blow up a rubber balloon to any desired size and shape. Wrap the wet yarn around the balloon in any design that your imagination dictates. Hang it up to dry. When the yarn is thoroughly dry, break the balloon. You will have a ball that is reminiscent of the string cages that used to hang in country stores. Glue a small figure to the bottom of the ball. A wooden toy angel was used in this one. String narrow ribbons through the ball, leaving short streamers at top and bottom. Add a sprig of mistletoe, hang it in a convenient place and watch the fun—or better still, participate in it.

A king-sized charm string as in Plate 89 makes a Christmas garland that will keep from year to year. Fresh evergreens may be used as a backing if desired, but there is charm in the simplicity and clear-cut outline of the cones and seed pods of this one. To make a charm string, cut a piece of medium weight rope any desired length. Wire each unit to be used separately and then wire them to the rope. Graduate the sizes. It is usually best to use the large ones at the top and the smallest ones at the bottom. The materials used in this one are all native to Oklahoma, gourds in variety—nest egg, birdhouse and dishrag gourds, okra, corn, chinaberry, yucca pods, cotton burrs, trumpet vine pods, honey locust beans and a few glycerin-treated magnolia leaves. It is well designed for the room and the place in which it is used.

For sheer fun and whimsy, try making a montage of old Christmas cards and scrapbook pictures. The one in Plate 90 is set in a deep Victorian frame that was painted gold and sprayed lightly

90. *Designed by:* Author   *Photographer:* Bryan Studio

with white paint, then decorated with snowflake sequins, Christmas ornaments and ribbon. It would be suitable to use in the hall or over the mantel of a house furnished with antiques, or it would make a delightful tongue-in-cheek decoration for a contemporary house. This is découpage, a medium little used in Christmas decorations but one with great possibilities. Découpage appeared as an "art" in the 18th century. Many since then have found fiendish joy in destroying pages, and creative satisfaction in pasting them together again, in designs never dreamed of before. All it takes is patience, a sense of design and imagination.

Browsing for material in antique shops, old book stores, and current magazines will bring you many happy hours. It is as much fun as any form of collecting. If you try it, let yourself go, express the real You with no inhibitions. Use tiny, sharp manicure scissors. The finer the cutting, the better the end result will be. Do not leave solid backgrounds.

Beautiful Christmas decorations can be made with straw. Its golden sheen combines well with bright red, green, turquoise and gold—all popular Christmas colors. It is a simple, natural medium suggestive of the humble birth of Christ. In the hands of an artist, as is evident in Plates 91, 92, and 93, straw can be fashioned into objects of sensitive beauty. Wheat straw may be purchased from

91. How To Make A Straw Star   92.
Designed by: Kathryn Holley Seibel   *Photographer: William Seibel*

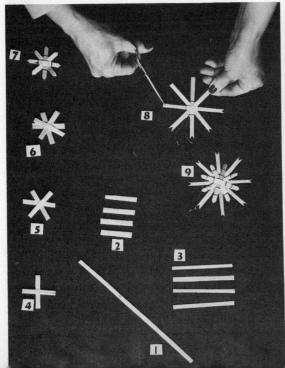

93. *Designed by:* Kathryn Holley Seibel
Photographer: William Seibel

a florist or better from a farmer at harvest time in June or July. Start with a simple unit, such as a star. Kathryn Holley Seibel who is well known for her beautiful work in straw sculpture, supplied the directions for making the star used on the Christmas tree in Plate 93. You will think of many more uses for this beautiful star.

Straw Designs and Directions

Preparation

Material for straw stars and snowflakes may be obtained from any available grain straw—wheat, rye or oat. Hand-cut straw is preferable; however a bale of straw will yield plenty of perfect pieces to use for small stars. Remove the husks from short pieces of straw and tie in bundles. Immerse in hot water for several hours or overnight before using.

Directions

(1) Flatten wet straw between thumb and forefinger.

(2) Cut four 2-inch pieces, taking care to match the widths of straw.

(3) Cut four 4 inch pieces, matching widths.

(4) Place two 2-inch pieces at right angles to form the small star part of 9.

(5) and (6) Place third and fourth 2-inch pieces at right angles, and superimpose on (4). Hold all together tightly.

(7) Hold end of 6-inch piece of thread tightly against the last straw (6) and weave thread in and out to catch and hold each piece of straw. Pull thread just tight enough to hold straws firmly in place and tie ends in a double knot. Trim ends uniformly into desired patterns. For variations see tree, Plate 93, and snowflakes, Plate 92.

(8) Repeat steps 4 to 7 with 4-inch pieces of straw to make large part shown in 9.

(9) Superimpose small star on large star and weave together. Tie thread securely. Press until dry.

Straw stars in variation and pieces of straw in abstract pattern decorate the unique tree in Plate 93. To duplicate it, cut a 42-inch cone from green suede paper, available at an art supply store, and glue the straw stars to it. A lamp shade covered with green suede paper forms the base.

94. *Wreath designed by:* Kathryn Holley Seibel
Photographer: Walter Bujak

In Plate 94, the wreath combines the golden seedheads of wheat with dried milkweed pods and artificial fruit. It is made on a 4-wire frame and a straw star is fastened to the center.

To make wreath, soak the wheat in hot water for one hour. Wire the heads together in groups of three. Floratape them to the outside and inside rim of the frame. Wire the milkweed pods to the second rim of the frame. Floratape the fruit to a single wire frame, superimpose it on the wreath and wire it to the frame.

See Plate 95 for a della Robbia snack basket used for candy, nuts or fruit. It makes a suitable table decoration from Thanksgiving through Christmas, with no upkeep or change of materials needed —a great convenience in this busy season.

To make it, use a small basket with a heavy rim that will take the stems of fruit. Cut six ¼-inch dowels 15 inches long. Floratape fruit and foliage to each dowel by the same method shown in the della Robbia wreath Plate 19. Use a tinker toy disc for the top and insert the finished sticks in the hole of the disc and in the rim of the basket at the bottom. Glue an angel to the top of the disc.

95. *Designed by:* Author   *Photographer:* Walter Bujak

The Christmas hat, vest and dog collars shown in Plates 96, 97 and 98 will add fun to any holiday party. They can be made long before Christmas and used for several years.

To make the gay Christmas hat of Plate 96, wrap a plastic hair clip with moss green ribbon. Sew treated holly leaves and artificial holly berries to it in a design to fit your hair style.

To duplicate the jolly vest for Dad, Plate 97, use a vest of one of his suits as a pattern and to give the correct size. Make a paper pattern first and then cut the two fronts from red felt. Cut Santa heads, candy canes, bells from white, green or blue felt. Decorate with sequins and sew to the vest. Cut two pieces of red grosgrain ribbon 2 inches wide in 27-inch lengths, sew to the vest to be tied at back. Put two buttons and buttonholes at the back of the collar to hold the vest in place.

97. *Designed by:* Mrs. Thomas I. Mairs
Photographer: Bryan Studio

96. *Designed by:* Mrs. Thomas I. Mairs
Photographer: Bryan Studio

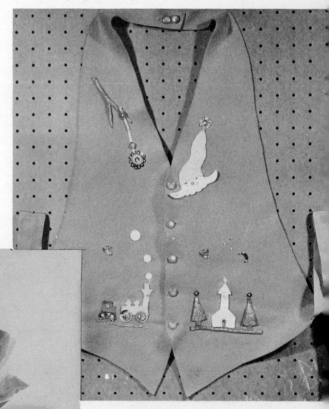

98. *Collars designed by:* Mrs. Thomas I. Mairs
Photographer: Bryan Studio

The delightful dog collars worn by the two toy poodles in Plate 98 can easily be made for your dog. Even the dog likes to think he is part of the Christmas fun. These two poodles enjoyed these collars and knew that fun was in store every time they were put on.

To make a collar, cut red felt 3 inches wide and the measurement of dog's neck. Decorate with sequins and red felt poinsettia bows. Use two tiny sleighbells for buttons.

All this and Flower Arrangements Too

Living plant material is a must in every gardener's and flower arranger's plan for Christmas decorating. The tinsel, glitter, gold paper and glass baubles have their place, but flowers satisfy the gardeners' need for the beauty of the earth.

Flower arrangements have always been needed. Through the ages they have been used for decorative purposes in most important celebrations. Records have come down to us through works of art from all the ages, even from eras before Christ was born, and each one expresses the spirit of the age in which it was made. This is a fact to ponder. It should inspire us to design arrangements that express the spirit of our modern age.

Flower arranging is an art, a very new one in our country, but one with great potentiality. We have only begun to be aware of its possibilities. Unfortunately, from the start, shackling manmade rules have hampered its progress; but happily we are now freeing ourselves from these rules and relying solely on the principles of design to guide us. Some may think that flower arranging contributes little or nothing to our culture, but this is not so. For thousands of people all over the United States, flower arranging satisfies an inherent need—a need to create beauty, and also a need for group expression and group participation. Many of us long to create beautiful music, to paint or write poetry, but most of us lack the training and the ability. With flower arranging, however, since most of the work has been done for us by the Great Designer, Who has given us the beautiful forms and colors to work with, anyone who will study the principles of design and how to apply them, can make beautiful arrangements. It is a most rewarding occupation, for it brings joy and satisfaction to the person doing it and pleasure to all who see her work.

Anything done right takes years of training and sensitivity. The longer we work the better we understand our medium and the more we enjoy it. The mature flower arranger discovers that it is not very exciting always to do patterns she knows, that for real enjoyment it is best to try new ideas, realizing that all art is a means of communicating feeling.

Inspired by the Christmas story and the spirit of Christmas, countless flower arrangements appear annually. Each one tells the story or a fragment of it in some special way. The successful ones convey the emotions of the artists who create them. A diversity of materials are used that often suggest the region where the arrangement was made. Each one radiates the spirit of Christmas.

Plates 99 to 117 are Christmas arrangements made for various types of homes in many different states. As you study these pictures you will be aware that it is not so much the material that is used as it is the spirit and the feeling of the person handling the material that conveys the true meaning and spirit of Christmas.

The journey of the Wise Men, bearing precious gifts for the Infant Jesus, inspired the beautifully designed arrangement in Plate 99. There is an emphasis on form and texture with a touch

99. *Designed by:* Mrs. W. H. Barton
Photographer: Gallap-Austin Studios

of elegance in the gilded containers (gold, frankincense, and myrrh) to suggest gifts fit for a King. Since the land in which Christ was born is in the same semitropical zone as Florida, plant material from Florida has been used. Branches and cones of tropical trees, magnolia leaves, seed pods, and a flower constructed of corn leaves, its edges touched with gold, used in the focal area, are combined to give the atmosphere of the Middle East. Touches of gold on the camel's accouterments repeat the gold of the three containers on the left, strengthening the horizontal balance.

Plate 100, an arrangement of elegance and beauty that complements the seventeenth century Dutch masterpiece by Abraham Bloomart, *Adoration of the Shepherd,* was exhibited in a Christmas show at the Atlanta Art Museum. Inspired by the painting "Christ, the same Yesterday, Today and Tomorrow," was the theme the arranger chose to interpret. Since the picture was Dutch, fruit was selected reminiscent of the Dutch fruit and flower paintings. It was sprayed lightly with gold paint, to make it symbolic of Today. The palm branches and bay leaves symbolic of Yesterday, were sprayed gold to tie in with the heavy gold frame of the picture. For Tomorrow, red anthuriums and a new golden fruit, a cross between the orange and tangerine, was used. Three flat, gilded containers were used, staggered to give depth to the design. Red, gold and blue repeat the medieval color harmony found in this old painting and in many of the Renaissance masterpieces depicting the Nativity.

100. *Designed by:* Marie Johnson Fort, (Mrs. Jesse)
Photographer: Dean King

As you study this composition you realize that the arranger has used design as a vehicle to express her creative powers. Having an idea is only a part of designing. Expressing the idea so that others will feel it is the major task.

We all have the ability to think and act creatively. All we need is experience. Have confidence and search for ideas within your own mind, rather than turning to experts.

In a sense, one lets oneself be creative by not being afraid of self, of others, or of the environment. Creativeness is a kind of by-product which comes without effort in people who have attained a healthy outlook and a keen awareness of life. Express your emotions freely in your work, become less inhibited. You will discover that creativeness will follow effortlessly, having some of the qualities of play.

The inspiration for the arrangement in Plate 101 came from the old French Christmas carol, *Sleep Little Jesus*—"Between the roses and the lilies there sleeps a little Child."

101. *Designed by:* Mrs. William A. Siemon
Photographer: William E. Buvinger

The container is an inverted 80-ounce champagne glass, obtainable from your florist. The Madonna and the Baby Jesus are modern porcelain. The arrangement is made of small artificial pink roses, lilies of the valley, blue forget-me-nots, and moss green velvet rose leaves. The leaf pattern on top of the glass is a complete gladiolus stalk, just as it grew with the flower stalk removed. The leaves were rolled up while green and fastened with Scotch tape and dried in an upright position. When dry the tape was removed. The leaves were placed in a needlepoint holder with pink strawflowers, gold, pink, and rose Christmas balls and moss green velvet rose leaves, then arranged on top of the inverted glass.

This is an artistic example of the new trend of combining artificial and natural material in flower arrangements. When well done, it can be beautiful as is evident in this composition.

For many years flower arrangers frowned on using artificial material. Happily we have now discovered that it has great possibiliites, is fun to work with, and has its place in flower arrangement. Perhaps, however, its place is not in the flower show, for there we emphasize the beauty and joy of gardening.

Keep in mind that it is not what you use but how you use material that helps or hinders a design. When you cling to the idea that only certain materials are right to use or that just certain materials belong together, you cut off a whole avenue of ideas and rob yourself of great possibilities.

For gardeners in the northern part of the United States, the Christmas rose (helleborus niger) is a favorite winter flower. In fact, it is about the only flower that can be found in northern gardens after the ground is covered with snow. This small flower loves the bitterness of winter and blooms underneath the snow. At Christmas time the snow is brushed aside, and the flowers are cut and used in very special arrangements to tell the Christmas story and delight the hearts of everyone who loves a garden. It is one of the oldest known flowers of our earth and yet surprisingly few people have ever seen it, or have known the delight of having this enchanting flower in their gardens. The artists who supplied these pictures have done much to revive interest in this almost forgotten flower by growing and supplying the plants to gardeners in many states. Many legends are told connecting the Christmas rose with the birth of Christ, making it a "natural" for a gardener's Christmas arrangement. The white starlike flowers that change to pink and green as they age, and the green seed pods that form later, blend perfectly with both conifer and broad-leaved evergreens. This easily grown plant will bring a new thrill to gardeners in all but our southernmost states.

In Plate 102, an angel discovers the beauty of the Christmas rose nestled against weathered wood on a base of rough sandstone. A fantasy of pure delight! The Christmas roses offer a pleasing contrast against the rough textured wood and stone. Observe how a very small amount of plant material has been used to convey great feeling and sensitivity.

102. *Designed by:* Mrs. Arthur Luedy
Photographer: Arthur E. Luedy

103. *Designed by:* Mrs. Arthur Luedy
Photographer: Arthur E. Luedy

Plate 103 illustrates a Christmas rose tree. A pair of these
would be beautiful on a dining table or mantel. The container
is an antique iron candlestick. Floral clay holds a cone of styrofoam to it. Twigs of Japanese holly, ilex crenata, often used in the
north instead of boxwood, are stuck into the cone. The Christmas
roses are put in water picks, available from a florist, and stuck
into the cone. The pleasure of producing a Christmas decoration
of living material from a northern garden is one that can be ex-
perienced by everyone who includes this beautiful plant in their
garden plan.

The very large Christmas decoration in Plate 104 was one of two unidentical arrangements that flanked a doorway in a Christmas show given by the Flower Arrangers Guild of Pittsburgh. The beautiful 12th century church figure dominates the composition. It is backed by a palm spray set in giant bamboo. Both are lightly sprayed with gold paint. Dignified and beautiful, it seems to suggest the great power of good that was set in motion at Christ's birth.

104. *Designed by:* Mrs. George Ketchum
Photographer: Brady Steward Studios

105. *Designed by:* Mrs. Raymond Russ Stoltz
Photographer: Arthur Heitzman

Plate 105 is an arrangement of strength and beauty, designed for a pine paneled library and placed on a handcarved, oak wine cabinet. The container is an antique copper water bottle from the Middle East. The evergreens and cones in large scale are right for the space they occupy. Artificial snow was sprayed on the material, heavy at the focal area to unify the composition with the parchment family crest on the wall.

106. *Designed by:* Marie Johnson Fort (Mrs. Jesse)
Photographer: Warbeck Studio

In Plate 106, an arrangement in the traditional Christmas red and green, was designed for a living room in Georgia. Burfordi holly with its own berries shading from the unripe green ones to the bright red ones, establishes the double crescent line of the design. Nandina berries at the base give weight and help to tie the plant material to the container. Red anthuriums were chosen because of their lasting quality and their color and placed to unify and balance the arrangement.

107. *Designed by:* Robert Rucker

Great feeling and sensitivity radiate from the arrangement in Plate 107. It is an excellent example of planned lighting to achieve a visual and photographic effect. A time exposure was taken with only the candles for light. The small, almost concealed candle in the foreground, casts a glow of light on the kneeling Mother and Child, giving a cathedral-like quality to the composition. A minimum of simple materials give a maximum of feeling and emotion.

108. *Designed by:* Mrs. Everett R. Combs
Photographer: The Towne House Studio

St. Anthony holding a figure of the Christ Child is featured in the arrangement, Plate 108. One of the happy memories of St. Anthony is the story of his setting up Christmas trees ladened with gifts of food for the poor. Sago palm, palm spathes, pine cones and Empress pawlonia seed pods used with copper glitter daisies are arranged on a walnut burl with the figure as part of the center of interest. The brown palm spathes are left natural and the other materials are sprayed with copper, relating it in color to the monk's robe and introducing a contrast of color and texture for variety and interest.

Plate 109 is a highly stylized version of the della Robbia theme in a Christmas arrangement. Beautifully designed with clear-cut interesting silhouette, it would add distinction to the holiday decoration of a contemporary house. This is an example of how the power of design declares a clear, direct and strong visual statement, complete as such.

109. *Designed by:* Mrs. Howard McClelland
Photographer: Ricci Studio

RICCI STUDIO

A contrast of textures and color spark the arrangement of white carnations and English holly foliage with bright red berries in Plate 110. The container is an alabaster bowl set in a brass holder. The use of elegant materials in formal balance was most appropriate for the French period living room where it was used.

110. *Designed by:* Mrs. Ernest E. Wunderly
Photographer: Bryan Studio

111. *Designed by:* Mrs. Edward Johnson
Photographer: Bryan Studio

Dried and treated plant material provide us with arrangements that last. When well designed as was the arrangement in Plate 111, they are a thing of beauty and a joy to live with.

Glycerin-treated leucothoe, loquat and magnolia leaves, and dried Scotch broom are used with deodara pine cones and peach stones wired and taped into a grape cluster. The alabaster container is heavily marked with brown tones and is placed on a brown stand to relate it to the plant material and complete the brown monotone color plan, sparked by the use of the light container.

112. *Designed by:* Mrs. Ernest E. Wunderly
Photographer: Bryan Studio

Plate 112 is designed for today and speaks for the Modern Age. The arranger has made dramatic use of a small quantity of plant material to produce an arrangement of importance that sparkles with the brilliance of Christmas.

Gilded strelitzia leaves are used with cream-white poinsettias that have beadlike gold centers. The brass container with black stand is placed on a marble coffee table, emphasizing the dramatic character of the composition.

113. *Designed by:* Mrs. Josiah O. Easterbrook
Photographer: Bryan Studio

114. *Designed by:* Mrs. Arnold Pariso
Photographer: Paul M. Clapper

Completely modern in material, design and feeling, Plate 113 depicts the eternal joy of motherhood. Here the power of form has been used to convey a modern feeling.

The arrangement in Plate 114 appeared in a Christmas show in Ohio and was entered in a class titled "O Holy Night." Thoughts of the Christmas Eve celebrations in our churches inspired this interpretation.

White plastic doilies sprinkled with multicolored glitter were glued to a white cardboard to represent the stained glass church windows. The edge of one of the doilies was glued to the rim of the cardboard as a frame. Glass tubes were painted gold and mounted in styrofoam to represent organ pipes. The choir boys stand in a loft of pine branches and cut pine cones that were sprayed gold. This arrangement could be used on a mantel or as a welcoming note on a hall table.

115. *Designed by:* Mrs. Paul L. Nimon
Photographer: Paul M. Clapper

Austrian pine sprayed with artificial snow and clipped to suggest a palm tree and branches, is used with a white ceramic Madonna in a modern Christmas arrangement, Plate 115. Striated plywood boards, lightly brushed with ice-blue flat paint, are used for the background. Ice-blue, pencil-slim tapers balance the design and carry the eye back to the center of interest.

117. *Designed by:* Mrs. John W. Knight, Jr.

116. *Designed by:* Mrs. Daniel H. Bishop
Photographer: Donald Porter

Plate 116, a sensitively beautiful interpretation of the spirit of Christmas, is done entirely with plant material native to Florida. The stylized, ice-blue Madonna set the color note for the arrangement. The niche behind the Madonna is a sheath of giant bamboo. It was sprayed with ice-blue paint in the same hue and value of the Madonna. The grayed blue-green of the eucalyptus leaves and seedpods complete a composition of simple dignity.

The candle-lighted antique brass lantern is the dominant feature of Plate 117. Using natural plant material only this arrangement is strong in design and symbolic of the light of knowledge that came to the earth with Christ's birth.

Index